Handbook of Colorimetry

HANDBOOK OF COLORIMETRY

PREPARED BY THE STAFF OF

THE COLOR MEASUREMENT LABORATORY

MASSACHUSETTS INSTITUTE OF TECHNOLOGY

Under the Direction of

ARTHUR C. HARDY

PROFESSOR OF OPTICS AND PHOTOGRAPHY
DEPARTMENT OF PHYSICS
MASSACHUSETTS INSTITUTE OF TECHNOLOGY

1936

THE TECHNOLOGY PRESS

MASSACHUSETTS INSTITUTE OF TECHNOLOGY

CAMBRIDGE, MASSACHUSETTS

Fifth printing, April 1970
Sixth printing, November 1973

ISBN 0 262 08001 X

Printed in
the United States of America

CONTENTS

LIST OF TABLES

PREFACE

DESPITE the age and extent of man's interest in color, colorimetry is a relatively new and unfamiliar science. Until physical instruments were developed which measure color in terms of quantities and wavelengths of light, the only available methods of color specification had of necessity to be based on samples of the various colors. The fact that these samples are subject to change with time, even under the best of conditions, has made it impossible to accumulate an extensive and accurate body of knowledge concerning the diverse phenomena of color and color vision. This handbook is concerned with the basis for the interpretation of the data obtained from physical measurements of colored materials, which may be expressed either in purely physical terms or in terms of the response of the normal observer as defined by the International Commission on Illumination in 1931. The preparation of this volume was undertaken by the staff of the Color Measurement Laboratory of the Massachusetts Institute of Technology as an essential part of its research program.

The first chapter, which is of an introductory nature, surveys the entire subject of colorimetry and has been written primarily for the benefit of those who are approaching this subject for the first time. This is followed by a discussion of the characteristics of light sources, the physical measurement of colored materials, and the laws of color mixture. The subsequent chapters include the recommendations of the International Commission on Illumination, which are here interpolated to wavelength intervals of one millimicron. These later chapters also include many auxiliary tables and charts that facilitate the specification of color.

It should perhaps be emphasized that this volume deals primarily with the subject of colorimetry. Little attempt has been made to treat the broader aspects of color phenomena for the reason that, as in all branches of science, methods of measurement must be standardized before the phenomena themselves can be investigated and intelligibly reported. By facilitating the measurement and specification of color, this volume should aid not only in fundamental research in this field but also in the industrial applications.

The members of the staff and student body who have assisted Professor Hardy in the preparation of this handbook are Dr. David L. MacAdam, Dr. Robert D. Nutting, Mr. Richard H. Zinszer, Mr. John Sterner, Mr. William E. Simmons, Mr. Laurence A. Horan, Mr. Lyman P. Hill, Mr. Harold R. Bellinson, Mr. Marvin I. Chodorow, Mr. Richard F. Morton, and Mr. D. G. Jelatis. We wish to acknowledge our indebtedness to Dr. Deane B. Judd of the Colorimetry Section of the National Bureau of Standards for many valuable suggestions and for permission to publish the results of some of his own calculations. We are also indebted to Dr. K. S. Gibson, Director of the Colorimetry Section of the National Bureau of Standards, and to Dr. A. W. Kenney of the E. I. duPont de Nemours & Company, Incorporated, for their kindness in reading and criticizing the galley proof. It is impossible to make adequate acknowledgment of the assistance and inspiration that have resulted from contacts with The Research Laboratories of The International Printing Ink Corporation where the concepts and methods described in this volume are finding daily application to the solution of practical color problems.

K. T. Compton

May, 1936

CHAPTER I

THE PHYSICAL BASIS OF COLOR SPECIFICATION

MAN has been conscious of color from the earliest times through the agency of the eye and its associated nervous system. There is ample evidence that even prehistoric man learned to decorate the walls of his cave with coloring materials that he dug from the earth. These mineral colors, a few vegetable dyes, and some substances obtained from certain insects and mollusks were the only materials available for many centuries. In fact, the paucity of suitable materials was not remedied until after Perkin's synthesis of mauve from coal tar in 1856. This discovery led to the production of literally thousands of new synthetic dyes and pigments. Despite the ever-increasing use of color during modern times, the lack of suitable coloring materials has ceased to be a serious problem.

As frequently happens, the solution of one problem created another. The synthesis of a vast number of new dyes and pigments made it necessary to look for a new method of color specification. When the number of available materials was small, it was feasible to designate the color of a dye or pigment by reference to its origin. Such terms as Tyrian purple, madder, henna, and indigo were used in this way. The modern equivalent of this procedure would be to give the chemical composition or structural formula of each new synthetic material, but such a specification would be inadequate for a number of reasons. To mention only one: dyes and pigments are generally used today in mixtures of two or more at a time. This is almost invariably a physical mixture rather than a chemical combination. Hence, the interpretation of mixture phenomena is to be found by application of the laws pertaining to the branch of physics known as optics, rather than by application of the laws of chemistry. The purpose of this volume is to indicate the physical method of color specification and to supply tables and charts which facilitate this specification.

1. Definition of Color

The term *color* is commonly used in three distinctly different senses. The chemist employs it as a generic term for dyes, pigments, and similar materials. The physicist, on the other hand, regards the term as a description of certain phenomena in the field of optics. Hence, the physicist, when confronted with the problem of measuring the color of a material, isolates and measures the relevant optical properties of the material. Physiologists and psychologists often employ the term in still another sense. They are interested primarily in understanding the nature of the visual process, and use the term on occasions to denote a sensation in the consciousness of a human observer. Color is a household word as well, and is commonly used indiscriminately in all three senses.

All three definitions of color are so firmly rooted in our language that it is futile to suggest that two of the meanings be abandoned in order to satisfy the scientist's desire for a single meaning for every term. Indeed, it would be difficult to obtain any degree of unanimity concerning the single meaning that should be adopted, even among scientists themselves. It seems inevitable that all three definitions of color will be continued, and that ambiguity can be avoided only by reference to the context. There is usually no difficulty in recognizing cases in which the chemical definition is intended. However, the distinction between the use of the objective physical definition and the subjective psychological definition is somewhat more subtle. This distinction can best be expressed in terms of some other concept such as temperature. Temperature may be defined objectively in terms of the expansion of mercury.[1] It may also be defined subjectively in terms of sensations. This analogue suggests at once that there is no one-to-one correspondence between the objective and subjective aspects. We know, for example, that, although the ocean may remain at the same temperature, as indicated objectively by a thermometer, it feels warmer on a cold day and colder on a warm day because it is contrasted with air at different temperatures. A hot or cold shower taken prior to entering the ocean has a similar effect. These phenomena have their counterparts in the field of color. The sensation that results when one looks at a colored object depends to a considerable extent upon the nature of the surrounding field and the nature of the field to which the observer has previously been exposed.

This volume is concerned chiefly with a fundamental specification for the color of objects or materials. A specification of this sort, which regards color as an inherent property of an object or material, must necessarily be based on objective measurements. Color in the sense that it is used throughout this volume may be defined explicitly in terms of a definite set of physical operations. Indeed, every genuine definition of any quantity is merely a statement of the operations by which the quantity is evaluated. Hence, the operations described in this chapter may be regarded as a definition of color.

2. The Spectrophotometric Specification of Color

The basis of the objective method of color specification will be clear from an outline of the procedure that is followed in the solution of a specific problem. Consider an object coated with a paint that would commonly be

[1] More fundamentally by means of the Kelvin thermodynamic scale.

called green. Let this test sample be illuminated by a suitable source of light. Imagine further that a prism is so placed as to disperse the light falling on the sample into its spectral components — violet, blue, green, yellow, orange, and red. Consider for a moment only a single component — the violet, for example. It is evident that a surface can not reflect more violet light than falls upon it. In other words, the reflection factor of the surface must have a value between 0 and 1. The exact value can be readily determined by means of an instrument known as a spectrophotometer;[2] and experiment shows that the value does not depend upon the intensity of the incident beam. Hence, the reflection factor of a surface for violet light is one of its inherent properties. The same argument applies with equal force to all the individual components of the spectrum. As a consequence, a purely objective specification for the color of a surface can be made in terms of the reflection factor for each spectral component. In the case of the example selected, the inherent color characteristics of the green paint might be indicated as follows:

Spectral Region	*Reflection Factor*
Violet	0.11
Blue	0.28
Green	0.33
Yellow	0.17
Orange	0.12
Red	0.06

The subdivision of the visible spectrum into six broad regions is an arbitrary procedure because the color of the spectrum varies continuously throughout its length. The physical difference between the various regions of the spectrum is merely one of wavelength, and the wavelength varies continuously from one end of the spectrum to the other. The unit of length that is commonly employed for specifying the wavelength of visible radiation is the millimicron (25,400,000 millimicrons equal one inch). In terms of this unit, the spectral regions mentioned above comprise approximately the following ranges in wavelength:

Violet	400–450 millimicrons
Blue	450–500 millimicrons
Green	500–570 millimicrons
Yellow	570–590 millimicrons
Orange	590–610 millimicrons
Red	610–700 millimicrons

Reflection factor measurements for six spectral regions, although useful as an illustration of the principle underlying spectrophotometric analysis, do not in general define the color of a reflecting surface with sufficient precision. Instead, the reflection factor is ordinarily determined for as many wavelength regions as the nature of the problem requires, and the results are usually expressed in the form of a chart like Fig. 1. The curve accurately defines the property of the sample that was roughly described by the data in the preceding tabulation. Obviously, every possible color can be represented on a chart of this type. For example, a perfectly white surface, which reflects completely all the visible radiations that fall upon it, would be represented by a horizontal line at the top of the chart. Similarly, an absolutely black surface would be represented by a horizontal line at the bottom.

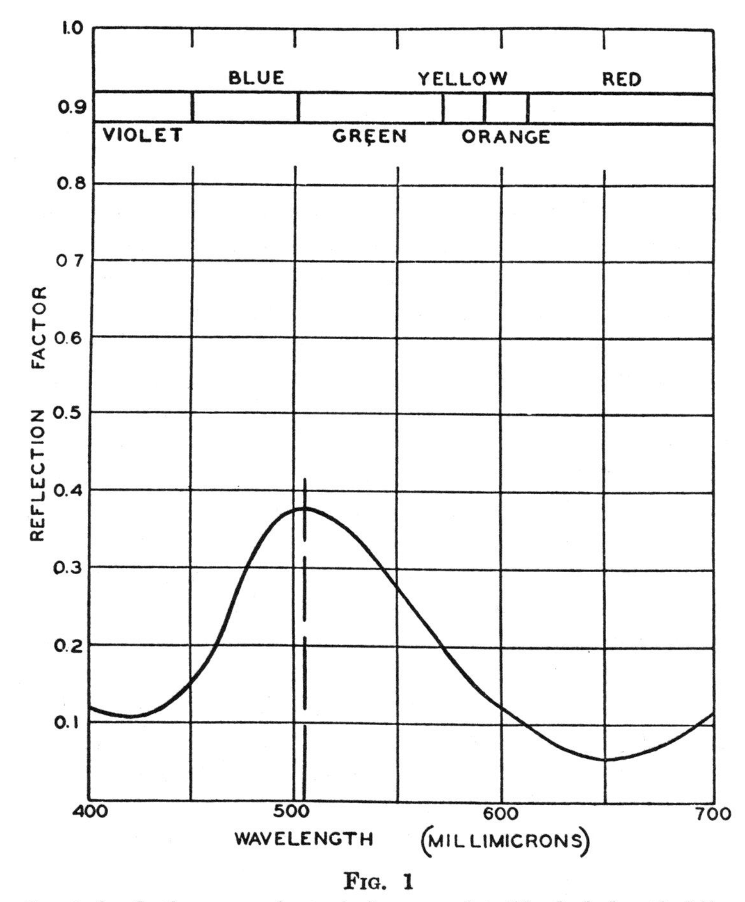

Fig. 1

Spectral reflection curve of a typical green paint. The dashed vertical line indicates the so-called dominant wavelength of this color. In this case its value is 506 millimicrons, which will be seen to correspond to a bluish green.

The spectral reflection curve of a material constitutes a permanent record that does not require the maintenance of a sample of the color. Furthermore, the units in which the curve is expressed are understood and accepted in every civilized country. The wavelength of light has been adopted internationally as the fundamental standard of length to which all other standards of length are referred. The values of the reflection factor are referred to a white standard whose factor is 1.00 for all wavelengths. Such a standard is achieved by absolute methods of reflectometry.[3]

Spectral reflection curves are at first somewhat confusing to those who are not acquainted with them, chiefly

[2] Chapter III is devoted largely to the subject of spectrophotometry.

[3] As a working standard, a surface of magnesium oxide is often used because it can be reproduced conveniently and with sufficient accuracy. It has a nearly constant reflection factor of more than 0.97 throughout the spectrum.

because the curves provide so much more information than can be obtained by visual examination.[4] The information supplied by the curve is essential in the solution of a vast number of problems, notably problems that are concerned with the obtaining of color matches that will be valid under any type of illumination, the maintenance of color standards, or the interpretation of mixture phenomena. However, there are other problems for which a knowledge of the stimulation of the eye is sufficient. A typical problem is the establishment of a color language conveying no more information than the eye can acquire by the ordinary visual examination. It will now be shown that the information supplied by a spectral reflection curve can be used as a basis for evaluating the stimulation that results under any set of specified conditions.

3. Standard Illuminants

The visual stimulation that results when one looks at a colored surface depends upon the character of the light by which the surface is illuminated. If all the energy radiated by the source of light is confined within a narrow band of wavelengths in the violet region of the spectrum, the surface will reflect only violet light. Stated more generally, if a surface is illuminated by light of substantially a single wavelength, it will reflect only light of this wavelength.[5] Hence, even though the green paint under consideration is known from the curve in Fig. 1 to reflect green light more effectively than light of other wavelengths, it may nevertheless be made to take on any hue of the spectrum if suitably illuminated. In view of this, it may be wondered how such a sample could have been identified with the green region of the spectrum before the advent of spectrophotometry. The reason is simply that daylight, which is the traditional source by which samples are examined, consists of a mixture of all the components of the visible spectrum in nearly equal proportions. Hence, when light of daylight quality falls upon the surface whose spectral reflection factor is indicated by the curve in Fig. 1, a preponderance of blue-green light is reflected into the eye of the observer.

The spectral reflection curves of a few typical colors will be found in Figs. 2 to 6. In each case, the wavelength of the spectral region with which each color is most closely identified when illuminated by daylight is indicated by the dashed vertical line. The curve for the

[4] The eye is, of course, incapable of analyzing light into its spectral components. For example, even such a common source as sunlight is not intuitively resolved into the various spectral colors that it contains. Those who are accustomed to mixing dyes and pigments sometimes claim the ability to see in the resulting mixture the components which they have added. This is merely a judgment based on experience and not analysis. It is interesting to note in this connection that the ear possesses the analytical power which the eye lacks. The ear is capable of analyzing as complex a stimulus as the music of a symphony orchestra into the components produced by the various instruments.

[5] This is not true of materials that exhibit fluorescence, but such materials are relatively rare.

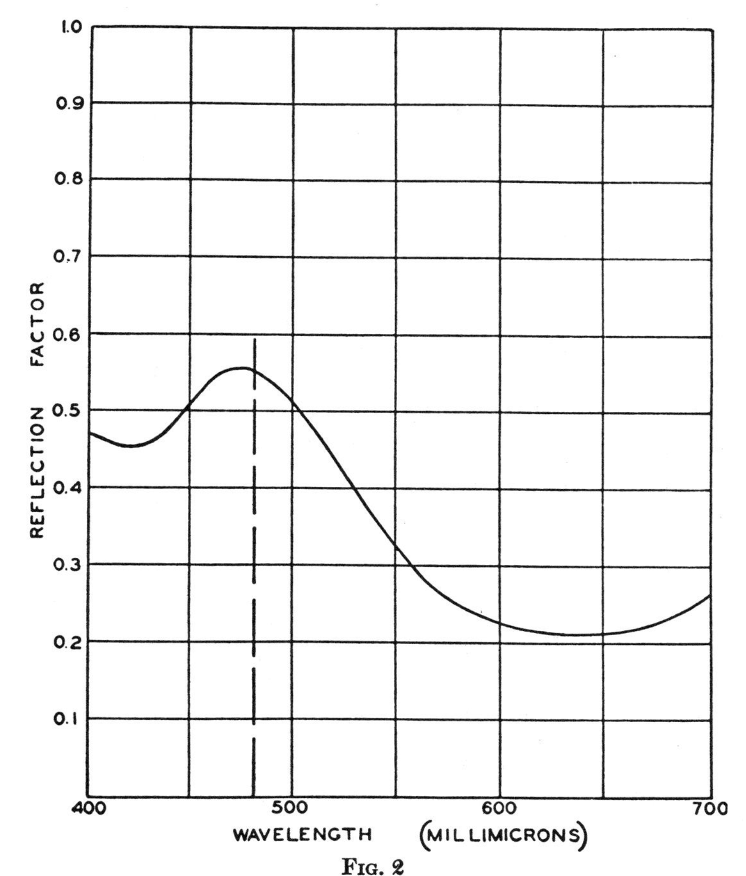

Fig. 2
Spectral reflection curve of a typical blue material with a dominant wavelength of 482 millimicrons.

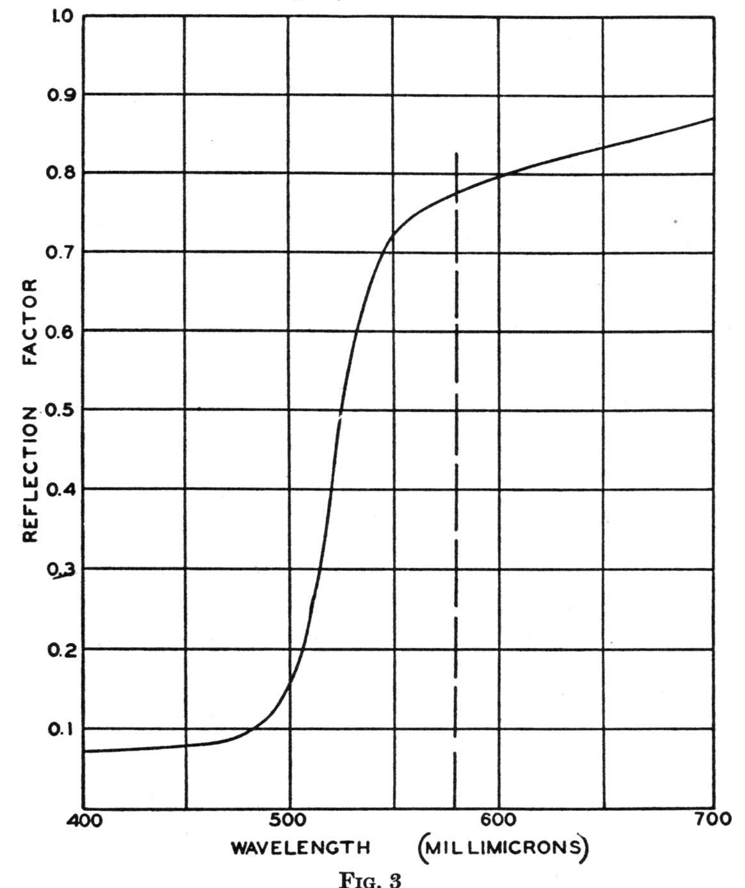

Fig. 3
Spectral reflection curve of a typical yellow material with a dominant wavelength of 579 millimicrons.

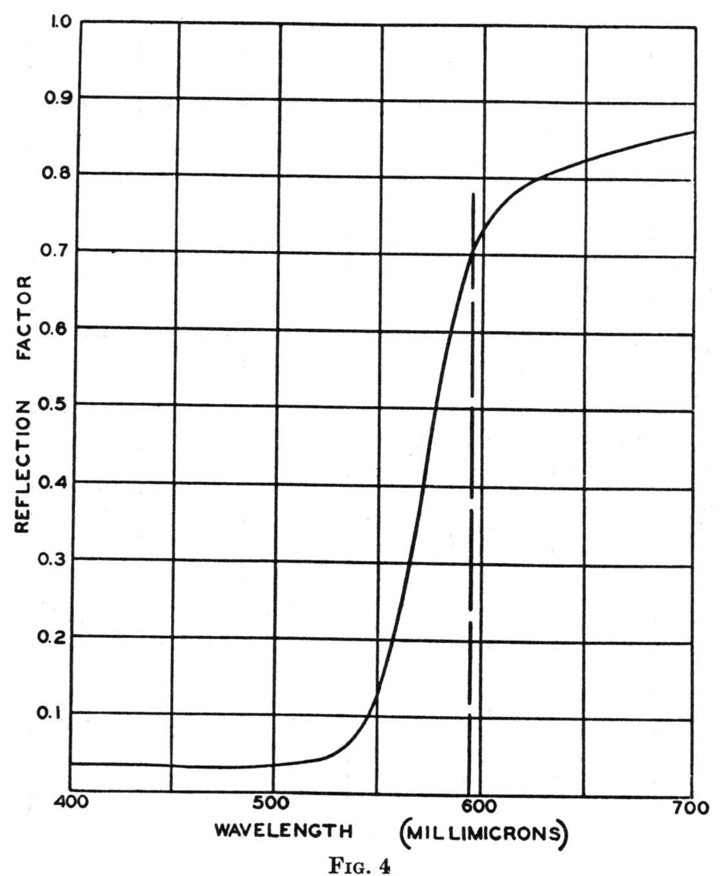

FIG. 4
Spectral reflection curve of a typical orange material with a dominant wavelength of 595 millimicrons.

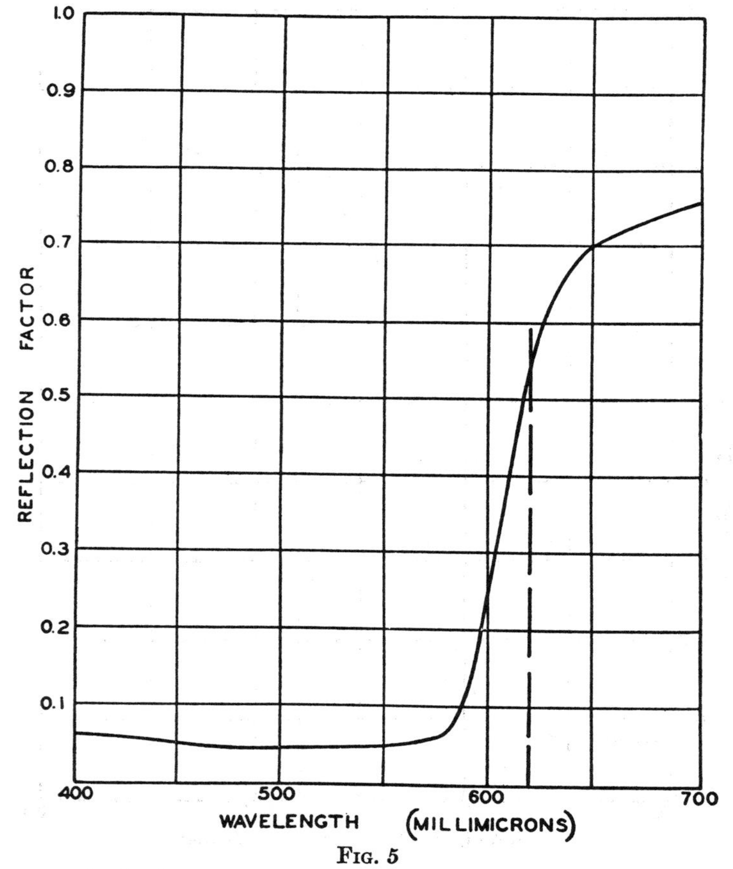

FIG. 5
Spectral reflection curve of a typical red material with a dominant wavelength of 619 millimicrons.

purple (Fig. 6) requires special mention because it possesses two regions where the reflection factor is high. Purple is not a spectral color but is a combination of red light and violet light. It is therefore impossible to find a single region of the spectrum that simulates purple, and the indicated dominant wavelength is that of the green that is complementary to it. For the present, the complement of a color may be considered to be the color that would produce a neutral gray in an additive [6] mixture. The spectral reflection curve of the complement is obtained by inverting Fig. 6.

The statement that daylight consists of a mixture of all the components of the spectrum in nearly equal proportions requires further elaboration. Extensive investigations into the distribution of energy in the spectrum of daylight have been carried on over long periods of time to average the effect of the state of the weather and the altitude of the sun.[7] A filter has been prepared which, when used with a tungsten lamp operated at the proper temperature, provides a source that is a close approximation to average daylight. The spectral distribution of energy in this source is shown by the curve in Fig. 7. At the meeting of the International Commission on Illumination in 1931, the representatives of the various countries adopted a source having this distribution of energy as an international standard of illumination to be used for the purposes of colorimetry except when special conditions dictate the use of other sources. This standard is known as I.C.I. Illuminant C.

Two other standards were adopted at the same meeting. They were designated as Illuminant A and Illuminant B respectively. The former represents a source having an energy distribution similar to that of a gas-filled tungsten lamp; the latter is an approximate representation of mean noon sunlight, and is slightly yellower than Illuminant C, being intermediate between Illuminants A and C. There is a strong argument in favor of one principal, internationally standardized illuminant representing daylight. Illuminant C is unquestionably the best representative of average daylight,[8] and it is being used to a much greater extent than Illuminants A and B. Although some data are given in later chapters of this volume for all three illuminants, the more extensive charts are all on the basis of Illuminant C.

[6] A mixture in which the light from each of the components reaches the eye in an unmodified state as is the case, for example, when the colors are combined on a Maxwell disc (see page 30). Mixtures of dyes or pigments are known as subtractive mixtures, for reasons that will be clear later.

[7] Briefly, the method employed in determining the spectral distribution of energy in a source of light consists in dispersing the light into a spectrum by means of a prism. Each spectral region is then isolated in turn, and the amount of energy present in each region is determined from the reading of a sensitive temperature-measuring device whose surface has been blackened so as to convert the light energy into heat.

[8] On a clear day, daylight is a mixture of sunlight and blue light from the sky. This mixture is found to have substantially the same quality as the light from any part of the sky on an overcast day.

If light having the spectral quality of Illuminant C falls on the surface of green paint whose spectral reflection characteristics are represented by the curve in Fig. 1, the spectral distribution of the energy reflected into the eye of an observer is obtained by multiplying at each wavelength the value given in Fig. 7 by the corresponding value in Fig. 1. The result of this operation is shown in Fig. 8. Because the curve in Fig. 7 is relatively flat, the shape of the new curve differs only slightly from that of Fig. 1. Of course, this is merely a special case which will serve to illustrate the principle that multiplying at each wavelength the incident energy by the reflection factor of the surface gives the distribution of energy in the light reflected by the surface. The next problem is evidently to determine how light with a known distribution of energy will stimulate the eye of an observer. No two observers respond in precisely the same manner, but the differences are remarkably small except in the case of those who can be definitely classed as color blind. The latter group, comprising only about two out of every hundred, will be ignored in this treatment of the subject.

4. Color Specification in Terms of Equivalent Stimuli

It is obviously impossible to expose an observer to light of a known spectral quality and expect him to describe the sensation that it produces. He could no more do this than he could describe the sensation produced by being pricked with a pin. Of course, he might reply in the latter case that being pricked with a pin evokes the same sensation as contact with a hot object or contact with a high tension wire. Such a reply is not a description of the sensation; but it does furnish the useful information that another stimulus evokes the same sensation. The analogy suggests the possibility of evaluating a color in terms of certain standard or primary stimuli. Indeed, it has been known for more than a hundred years that a normal observer can duplicate the effect of any color stimulus by mixing the light from three primary sources in the proper proportions. A simple experimental technique by which this may be accomplished in the vast majority of cases is as follows: The observer looks into an optical instrument containing a suitable photometric field. The light whose color is to be matched is introduced into one half of the field, and the light from the three primary sources is introduced in controlled amounts into the other half. By manipulation of the controls, a setting can be found where an exact color match between the two halves of the field is obtained. Furthermore, there is only one setting for each of the three controls that will produce a color match. By calibrating the controls, the amount of each primary can be recorded. The unknown color can then be specified by three numbers, X, Y, Z. These are known as the tristimulus values, each number representing the amount of one of the primary stimuli.

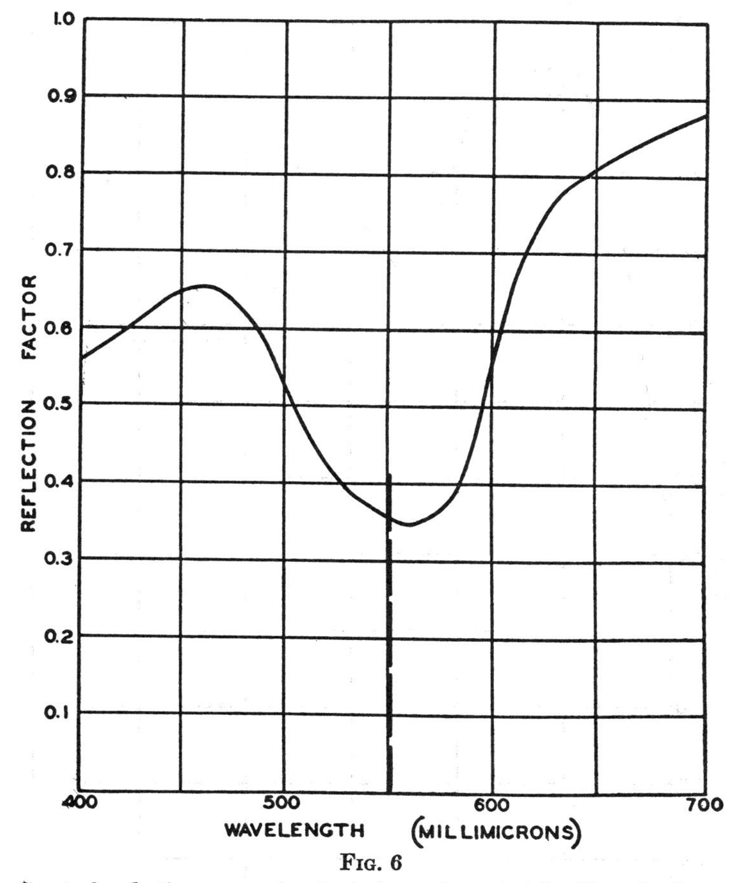

Fig. 6

Spectral reflection curve of a typical purple material with a dominant wavelength of 550c millimicrons.

Instruments of this type which synthesize an equivalent stimulus are known as colorimeters. The tristimulus values obtained with a properly designed colorimeter constitute a satisfactory color specification for a given test sample and for the observer who determines the equivalent stimulus. If another observer performs the same experiment, he will obtain an adequate specification of the color in terms of the stimulus which he regards as equivalent. The two specifications may be slightly different, even though neither observer could be classed as color blind. Consequently for inter-laboratory comparisons or for long-time color standardization programs based on the use of colorimeters, a large group of observers must be employed. An alternative procedure has been devised and has been recommended for international use. This procedure consists in determining certain basic color mixture data for a large group of carefully selected observers. These basic data can then be used in conjunction with spectrophotometric data to compute for any test sample the average tristimulus values that would have been obtained by this group of observers if they had used a colorimeter. Since the readings obtained with a spectrophotometer are independent of peculiarities of an observer's eye, this procedure provides a basis for the specification of color in terms of the average chromatic properties of an internationally accepted group of observers.

The basic data required for the computation of such a

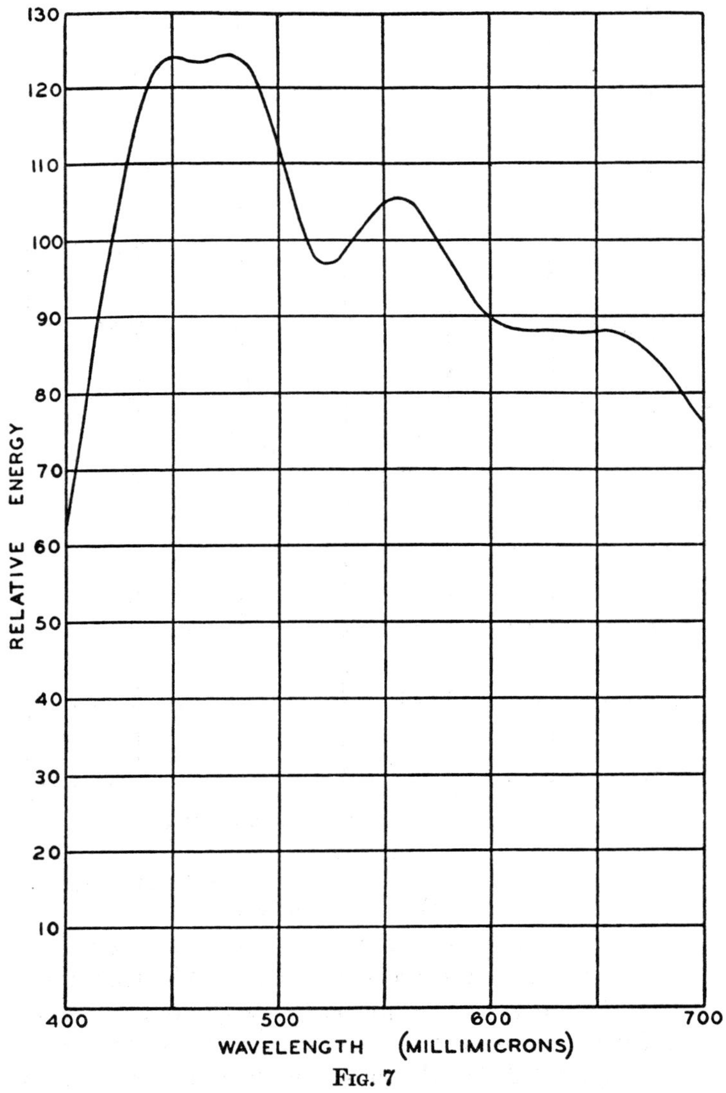

FIG. 7

Relative spectral distribution of the energy radiated per unit time by I.C.I. Illuminant C.

result must necessarily be obtained by the use of a colorimeter. Since these data need be determined only once, the observance of the various precautions associated with colorimetry is not a serious obstacle. The procedure by which these basic data are determined is substantially as follows: One half of the photometric field of a colorimeter is illuminated by a measured quantity of light of approximately a single wavelength — 400 millimicrons, for example. An observer is then asked to determine the amount of each of three primaries required to color match both halves of the field. These amounts of the primaries are the tristimulus values for this quantity of light of this wavelength. The wavelength of the light is then changed, say to 410 millimicrons. Again a color match is made and the tristimulus values are recorded. This process is continued until the entire visible spectrum has been examined. In the report of the data, adjustment is made for the amount of energy employed in each determination.

Investigations of this sort were carried out by Maxwell (1854), by König and Dieterici (1892), and by Abney (1913). The results were chiefly of academic interest until 1922 when the Colorimetry Committee of the Optical Society of America summarized and republished them in a convenient form. The publication of these O.S.A. data was an important event in the development of the science of colorimetrics. Before 1922, spectrophotometers had been used when the nature of the problem demanded a wavelength by wavelength analysis, and colorimeters had been used when only the evaluation of an equivalent stimulus was required. After the publication of these data, it became feasible to base tristimulus specifications on spectrophotometric data. A spectrophotometer can now be made to do the work of two instruments, and the uncertainties associated with colorimetry can be avoided.

Unfortunately the O.S.A. data were based on work that had been done at a time when the experimental facilities were none too adequate. However, about 1928 Wright and Guild in England independently redetermined these fundamental data, each employing a number of carefully selected observers. When their results were reduced to a comparable basis, it was discovered that their data were in extraordinarily good agreement, and were also in remarkably good agreement with the O.S.A. data. Because the subject had by this time assumed considerable importance, the International Commission on Illumination undertook to bring about an international standardization, which was consummated at the 1931 meeting. It is almost certain that these I.C.I. recommendations will not be altered for many years to come. The experimental technique employed in the determination of the data was excellent, and the number of observers was sufficient to justify the belief that a greater number would not make significant changes in the average.

In the experiments conducted by Wright, the primaries employed were spectrum colors whose wavelengths were 650, 530, and 460 millimicrons respectively. In Guild's experiments, on the other hand, the primaries were produced by means of red, green, and blue filters. Each investigator determined the tristimulus values for the various spectrum colors in terms of the set of primaries that he had adopted. Since Wright and Guild used different primaries, they obtained different tristimulus values. For the purposes of colorimetry, however, either set of data is adequate, because the primaries are employed merely as intermediates in terms of which two colors can be compared. On this basis, any set of primaries is adequate, and the advantage of one set over any other set reduces merely to a matter of convenience.

There is one disadvantage common to the primaries used by both Wright and Guild. When matching the various spectrum colors, both investigators found it necessary to use negative amounts [9] of at least one of the primaries. Hence, the results are in an inconvenient form

[9] A negative amount of a primary is obtained, in effect, by transferring the primary to the opposite side of the field where it is combined with the radiation whose tristimulus values are being determined.

for use in computation because of the presence of both positive and negative values. It will be clear later that no set of real primaries can be found that will match all colors (including the spectrum colors) without employing negative amounts of at least one of the primaries. It follows that, if negative tristimulus values are to be avoided, the primaries must be chosen outside the realm of real colors. Fortunately this is no hardship because, when the basic data have once been determined for one set of primaries, the results that would have been obtained with any other set can easily be calculated by a simple linear transformation.[10]

The tristimulus values that were adopted by the International Commission on Illumination for the various spectrum colors are given in abridged form in Table I and are represented graphically in Fig. 9. The

TABLE I

Tristimulus Values of the Spectrum Colors

Wavelength	$\bar{x}$	$\bar{y}$	$\bar{z}$
400	0.0143	0.0004	0.0679
410	0.0435	0.0012	0.2074
420	0.1344	0.0040	0.6456
430	0.2839	0.0116	1.3856
440	0.3483	0.0230	1.7471
450	0.3362	0.0380	1.7721
460	0.2908	0.0600	1.6692
470	0.1954	0.0910	1.2876
480	0.0956	0.1390	0.8130
490	0.0320	0.2080	0.4652
500	0.0049	0.3230	0.2720
510	0.0093	0.5030	0.1582
520	0.0633	0.7100	0.0782
530	0.1655	0.8620	0.0422
540	0.2904	0.9540	0.0203
550	0.4334	0.9950	0.0087
560	0.5945	0.9950	0.0039
570	0.7621	0.9520	0.0021
580	0.9163	0.8700	0.0017
590	1.0263	0.7570	0.0011
600	1.0622	0.6310	0.0008
610	1.0026	0.5030	0.0003
620	0.8544	0.3810	0.0002
630	0.6424	0.2650	0.0000
640	0.4479	0.1750	0.0000
650	0.2835	0.1070	0.0000
660	0.1649	0.0610	0.0000
670	0.0874	0.0320	0.0000
680	0.0468	0.0170	0.0000
690	0.0227	0.0082	0.0000
700	0.0114	0.0041	0.0000

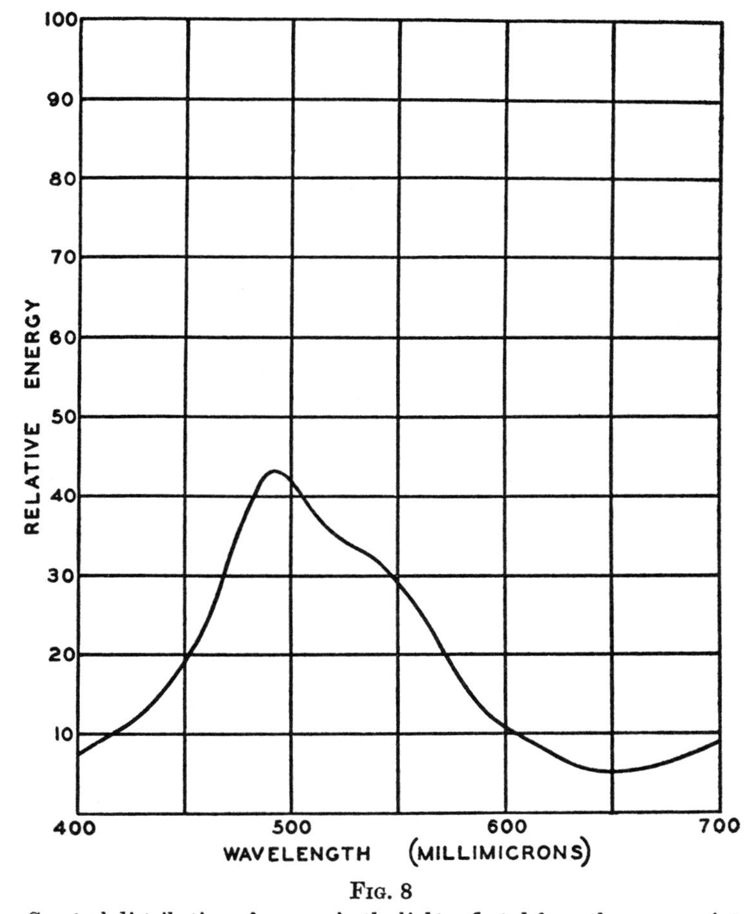

Fig. 8

Spectral distribution of energy in the light reflected from the green paint of Fig. 1 when the illuminant has the energy distribution of Fig. 7.

values of $\bar{x}$, $\bar{y}$, $\bar{z}$ in this table indicate the amount of each of the I.C.I. primaries that is required to color match a unit quantity of radiant energy of the various wavelengths. The fact that none of these values is negative results from a transformation of the data of both Wright and Guild to a set of primaries lying outside the realm of real colors. Although, as mentioned before, the choice of primaries is quite immaterial, it may assist the reader in the interpretation of the I.C.I. tristimulus values to have in mind some concept of the nature of these primaries. The value of X (or $\bar{x}$ in the case of spectrum colors) represents the amount of a primary which is a reddish purple of higher saturation than any obtainable color having this hue. The value of Y (or $\bar{y}$) represents the amount of a green primary considerably more saturated than the spectrum color whose wavelength is 520 millimicrons. The value of Z (or $\bar{z}$) represents the amount of a blue primary that is considerably more saturated than the spectrum color whose wavelength is 477 millimicrons.

The necessary data are now available for evaluating a stimulus that is equivalent to the sample of green paint when the latter is illuminated by light having the spectral quality of Illuminant C and is viewed by the standard

[10] This transformation is of the form

$$r' = k_1 r + k_2 g + k_3 b \quad g' = k_4 r + k_5 g + k_6 b \quad b' = k_7 r + k_8 g + k_9 b$$

where r, g, and b are tristimulus values based on one set of primaries, r′, g′, b′ are tristimulus values based on the new set of primaries, and $k_1, k_2 \ldots k_9$ are the tristimulus values of the original primaries on the basis of the new primaries.

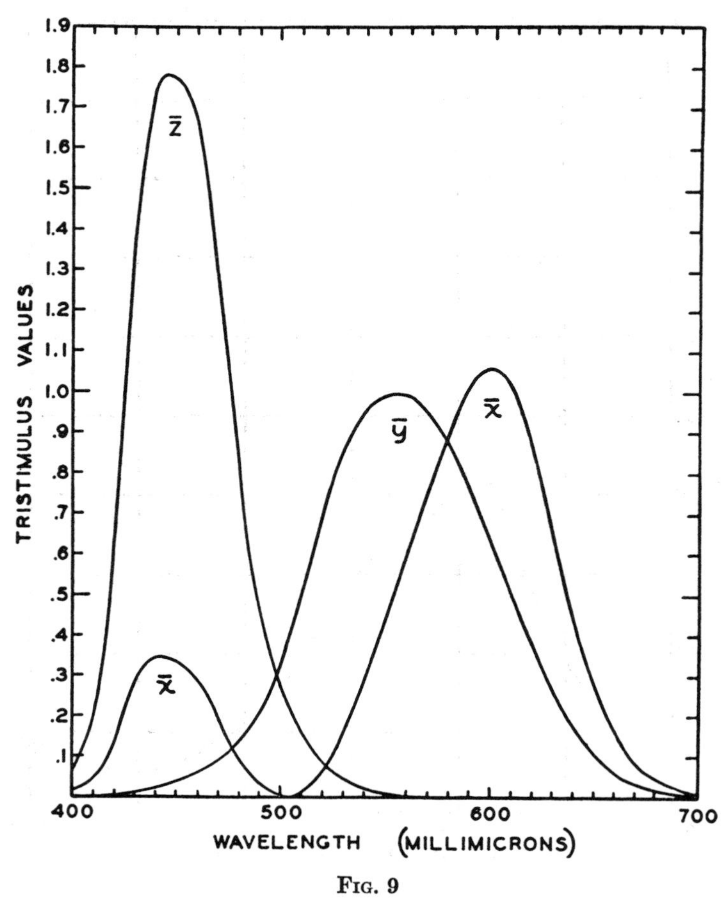

Fig. 9

Tristimulus values for the various spectrum colors. The values of $\bar{x}$, $\bar{y}$, $\bar{z}$ are the amounts of the three I.C.I. primaries required to color match a unit amount of energy having the indicated wavelength.

I.C.I. observer. The details of the procedure by which this result is computed will be discussed and illustrated by examples in later chapters. In principle, the method consists in dividing the visible spectrum into a suitable number of equal wavelength intervals, determining the contribution to the stimulation made by the light within each interval, and then summing the result. In the case of the green paint selected for illustration, the tristimulus values that result from such a computational procedure are $X = 15.50$, $Y = 24.19$ and $Z = 22.64$. These values are a measure of the stimulation in the sense that each number indicates the amount of one of the three primary stimuli that would be required for a color match if the I.C.I. observer were to use a colorimeter employing the transformed I.C.I. primaries.[11]

5. Two Methods of Color Matching

It must be emphasized again that these tristimulus values represent a stimulus that is equivalent to the test sample only when the illumination of the sample is by light having the spectral quality of Illuminant C, the sample is observed by the standard I.C.I. observer, and the observing conditions are like those used in determining the basic I.C.I. data. Everyone knows that a green appears bluer on a yellow background and yellower on a blue background. Phenomena of this sort are important to artists and designers and also to those who attempt to understand the nature of the visual process. In the vast majority of instances, however, the chief problem is to know how to indicate that a sample from one lot of material can be used interchangeably with a sample from another lot. If the tristimulus values are the same in the two cases, the two samples will be regarded as a match by a normal observer in the sense that they may be used interchangeably under light having the spectral quality of Illuminant C. If the additional condition is imposed that the two samples must match for all observers under all conditions of illumination (including such extreme cases as illumination by light of a single wavelength), the two samples must have identical spectral reflection characteristics.[12] In general, a physical match of this sort requires the use of the same dyes or pigments. It is not always safe to assume that dyes or pigments that are sold under the same name are identical in either a physical or chemical sense. The manufacturers of dyes and pigments commonly blend several batches of material to produce a visual color match with a standard. This may not be a physical match as can easily be ascertained by spectrophotometric analysis.

[11] The computational procedure that is indicated here is unnecessarily tedious. Several modifications which result in simplifying the calculations will be discussed in later chapters. In fact, it is possible to modify an ordinary calculating machine in such a manner that these calculations can be performed automatically.

Attempts have been made to design photoelectric colorimeters for the direct determination of tristimulus values. In these instruments, the response of a photoelectric cell to the light reflected from the sample is recorded when each of three suitably selected filters is placed in the beam. It is obvious that the spectral transmission characteristics of the filters must be so chosen with respect to the energy distribution of the source and the spectral sensitivity of the cell that tristimulus values are either indicated directly or can be derived by a simple transformation. No instrument satisfying these conditions has yet been constructed. In fact, even if such an instrument were constructed, it would be useful as a colorimeter only so long as the spectral characteristics of the light source, cell, and filter, remained unchanged.

An instrument consisting of a photoelectric detector and a set of filters may be used for certain purposes, even though it may not be properly termed a colorimeter. For example, there are many problems in which a sample is to be compared with a standard having spectral reflection characteristics that are very similar. In this case, the filters should ordinarily be selected in such a manner as to isolate the region or regions of the spectrum where the departure of the sample from the standard is most likely to occur.

[12] These statements are rigorously true if the two samples under consideration are alike in all other respects. They must, of course, have identical surface structures and identical gloss characteristics. This is the usual situation. The problem occasionally arises of color-matching materials which are widely different in some of their other characteristics. For example, it is sometimes desired to match a sample of satin with a sample of crêpe. It is impossible to dye these materials in such a manner that they could be used interchangeably. The best that can be done is to secure a color match for some mode of illumination and observation. Experience has shown that samples of satin and crêpe which have identical tristimulus values for diffuse illumination and normal observation are regarded as a commercial color match. (See page 28.)

6. Need for a Universal Color Language

Students of history agree that man's progress was slow until he had developed a language that enabled him to impart to others the experience that he had acquired. Until the development of spectrophotometry, there was no satisfactory basis upon which to build a fundamental language of color. Because spectrophotometers have been developed quite recently in comparison with the age of the art, the existence of the instrument and the concepts associated with it are still unknown even to many who deal with color as a matter of daily routine. However, the spectrophotometric method is so straightforward that, as soon as one becomes familiar with the concepts, he finds it virtually impossible to think in any other terms. In fact, until one has sensed the possibility of a universal and fundamental language of color, he is not likely to realize the handicap caused by the lack of such a language. It does not strike him as incongruous that, when purchasing almost any article of commerce, he can read and understand the specification of all its important physical characteristics (such as size and weight) but is unable to acquire even an approximate idea of its color without seeing a sample. To construct a color language on the basis of samples of colored materials is not a fundamental solution to the problem because the samples are certain to be of questionable permanence and difficult to reproduce. On the other hand, spectrophotometry depends only on measurements of the wavelength of light and measurements of reflection (or transmission) factors, both of which can be determined with more than adequate precision. Although the tristimulus values do not provide so much information as the spectrophotometric data, they can be derived from the spectrophotometric data by a straightforward computation procedure. They therefore provide a fundamental basis for a language of color.

7. Brightness

It has been shown that if two materials can be represented by identical tristimulus values, they will constitute a color match under the standard conditions. It is unfortunate that the tristimulus values do not indicate in a readily comprehensible manner the nature of the color difference when a difference exists. In a simple case, if two colors are represented by $X = 40$, $Y = 50$, $Z = 30$ and $X' = 20$, $Y' = 25$, and $Z' = 15$ respectively, it can be reasoned that the two colors are alike in quality or chromaticity, but that one is twice as bright as the other in the sense that it reflects twice as much light. If the chromaticities had been different, the comparison of brightness would have been more difficult but for the foresight employed at the time the basic I.C.I. data were adopted. Long before the 1931 meeting of the International Commission on Illumination, there had been a universal agreement concerning the relative brightness

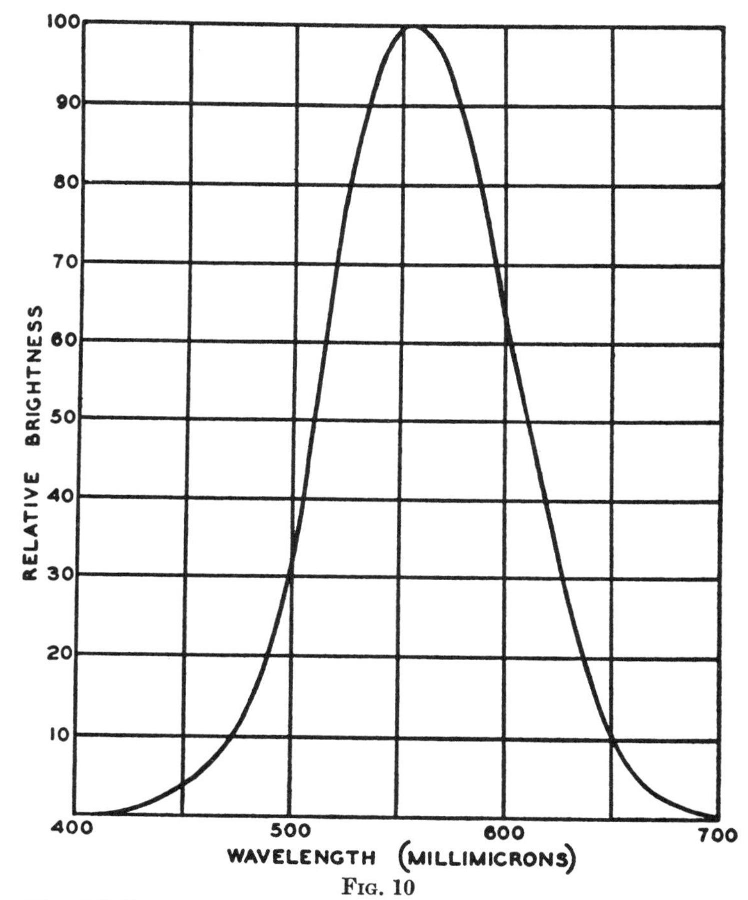

Fig. 10

The visibility curve for a normal eye. Specifically, this curve indicates the relative brightness of equal amounts (energy basis) of spectrum colors of the various wavelengths. The maximum visibility occurs at 555 millimicrons, and the visibility becomes substantially zero at 400 and 700 millimicrons.

of equal amounts of energy at various wavelengths. These data are presented in Fig. 10. In transforming the data obtained with the actual primaries used in the experiments of Wright and Guild to a set of primaries that would avoid negative tristimulus values, a set of primaries was chosen which made one of the three functions (the $\bar{y}$ function given on page 7) correspond exactly with the so-called visibility function shown in Fig. 10. Hence, the relative brightness of a sample is indicated directly by the value of Y on a scale that represents an absolute black by zero and a perfect white by 100. Thus the green whose tristimulus values are given on page 8 has a brightness relative to the white standard of 24.19%.

8. Trichromatic Coefficients

The evaluation of the quality of a color (chromaticity) is accomplished by defining three new quantities as follows:

$$x = \frac{X}{X + Y + Z} \tag{1a}$$

$$y = \frac{Y}{X + Y + Z} \tag{1b}$$

$$z = \frac{Z}{X + Y + Z} \tag{1c}$$

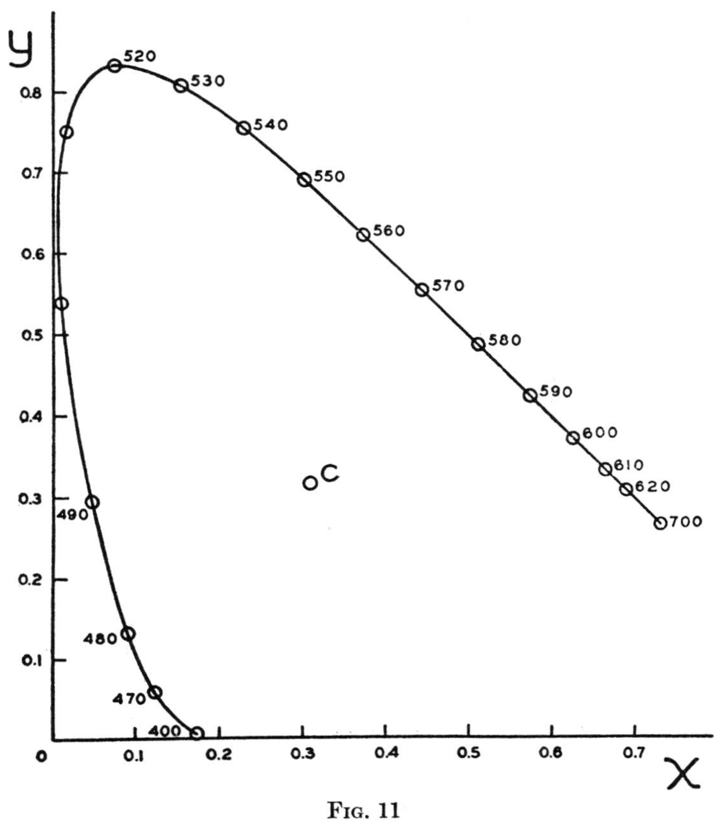

FIG. 11

Chromaticity diagram showing the locus of the spectrum colors and also the location of Illuminant C.

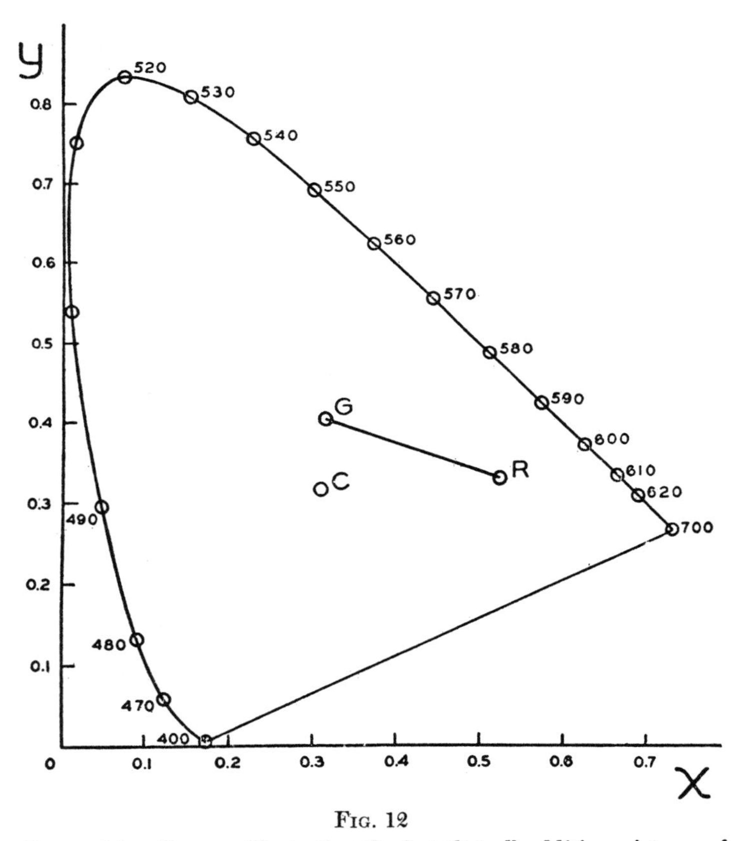

FIG. 12

Chromaticity diagram illustrating the fact that all additive mixtures of the colors R and G lie on the line RG. A similar consideration shows that all mixtures of the spectrum colors lie within the area enclosed by the solid line. Hence this area contains all the real colors.

These quantities are called trichromatic coefficients, although trichromatic coördinates might be a better term for reasons that will appear later. It will be noted that only two of these quantities are independent since

$$x + y + z = 1$$

regardless of the values assigned to X, Y, and Z. Hence to specify the chromaticity of a sample, it is necessary to give the values of only two of the three quantities (x and y have generally been selected for this purpose). Evidently, instead of using the tristimulus method of color specification, even more information is supplied if the color is specified in terms of Y, x, and y. As a practical example, compare the color considered on page 8 specified by $X = 15.50$, $Y \doteq 24.19$, $Z = 22.64$ with another for which $X' = 25.26$, $Y' = 39.42$, $Z' = 36.89$. When these are specified in the new notation they become respectively $Y = 24.19$, $x = 0.2487$, $y = 0.3881$ and $Y' = 39.42$, $x' = 0.2487$, $y' = 0.3881$. It is then apparent that the two colors have the same chromaticity, the difference between them being merely a brightness difference. This relationship would not have been immediately evident from the tristimulus values.

9. GRAPHICAL REPRESENTATION OF CHROMATICITY

For the same reason that it is impossible to learn geography without the use of maps, it is difficult to understand the subject of color without some form of graphical representation of the relationship of the various colors one to another. To represent the tristimulus values graphically would require a three-dimensional coördinate system which is obviously impracticable. The chromaticity can be conveniently represented, however, merely by plotting the trichromatic coefficients, x and y, on a two-dimensional diagram. This has been done in Fig. 11 for certain colors of outstanding interest. The solid line represents the locus of all the spectrum colors. This line is determined by computing the trichromatic coefficients of each of the spectrum colors from the tristimulus values that were given in the table on page 7. Thus, in the case of radiation having a wavelength of 600 millimicrons, the tristimulus values are $\bar{x} = 1.0622$, $\bar{y} = 0.6310$, and $\bar{z} = 0.0008$. From these it can be calculated that the trichromatic coefficients are $x = 0.6270$ and $y = 0.3725$. The tristimulus values of Illuminant C can be determined by an integration process (discussed in Chapter V) to be $X = 98.04$, $Y = 100.00$, and $Z = 118.12$. The corresponding trichromatic coefficients, $x = 0.3101$ and $y = 0.3163$, locate the position of Illuminant C.

A graphical representation like Fig. 11 is often called a color diagram but is preferably called a chromaticity diagram. Certain early attempts to represent colors graphically employed a trilinear coördinate system rather than the cartesian system that is used in Fig. 11. In the trilinear coördinate systems, the primary stimuli were placed at the apices of an equilateral triangle, and the diagram was called a color triangle. The term still persists although it is not so significant in connection with the I.C.I. coördinate system.[13]

10. Dominant Wavelength and Purity

A chromaticity diagram has one property that makes it of immense value in connection with the additive mixture of two or more colors.[14] In Fig. 12, suppose that a certain red is located at R and a certain green at G. Regardless of the proportions in which these colors are additively mixed, the resultant color will always lie on the line joining R and G, the exact position of the point being determinable by a procedure that will be discussed in Chapter IV. Because of this additive property of the diagram, it will be seen that all real colors must lie within the area enclosed by the solid line, since every real color can be considered to be a mixture of its spectral components in various proportions. Furthermore, all the real colors lying within the solid line but above the dashed line in Fig. 13 can be considered to be a mixture of Illuminant C and spectrum light of a certain wavelength. For example, the green G whose tristimulus values were previously computed is shown to be a mixture of Illuminant C and spectrum light having a wavelength of 506 millimicrons. This wavelength is known as the dominant wavelength.[15] Since this green lies on a line which terminates at a pure spectrum color at one end and at the illuminant point at the other end, the sample is evidently not so pure a green as the corresponding spectrum color. A numerical specification for the purity of this sample can be achieved by merely determining on the chromaticity diagram the relative distances of the sample point and the corresponding spectrum point from the illuminant point. In this case, the distance of the sample point from the illuminant point is 20% of the distance of the spectrum locus from the illuminant point. The sample is therefore said to have a purity of 20%.

The portions of the diagram lying within the solid line but below the dashed line in Fig. 13 represent the purples or magentas. It was previously implied and is now evident that purple cannot be obtained by mixing white light with a single spectrum color. Hence, an artifice is necessary if the concepts of dominant wavelength and purity are to be used in the specification of purples. The artifice consists in making use of a principle, which

[13] In the I.C.I. coördinate system, the trichromatic coefficients of the primaries are $x = 1, y = 0$; $x = 0, y = 1$; and $x = 0, y = 0$.

[14] This property of the diagram follows directly from the fact that the tristimulus values of an additive mixture are the sums of the tristimulus values of the components. Various types of additive mixtures are discussed in Section 37 of Chapter IV. The tristimulus specification is inadequate for the calculation of the result of subtractive mixtures, such as mixtures of dyes and pigments, and resort must be had in this case to spectrophotometric data.

[15] This quantity was referred to previously without definition in connection with Figs. 1 to 6.

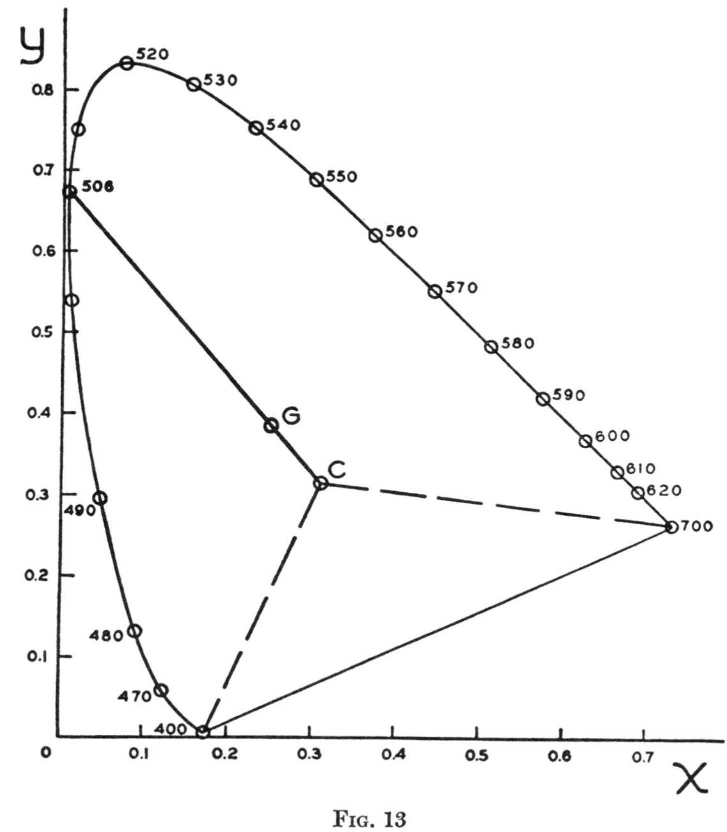

Fig. 13

Chromaticity diagram showing that the green (G) may be regarded as a mixture of Illuminant C and a spectrum color having a wavelength of 506 millimicrons.

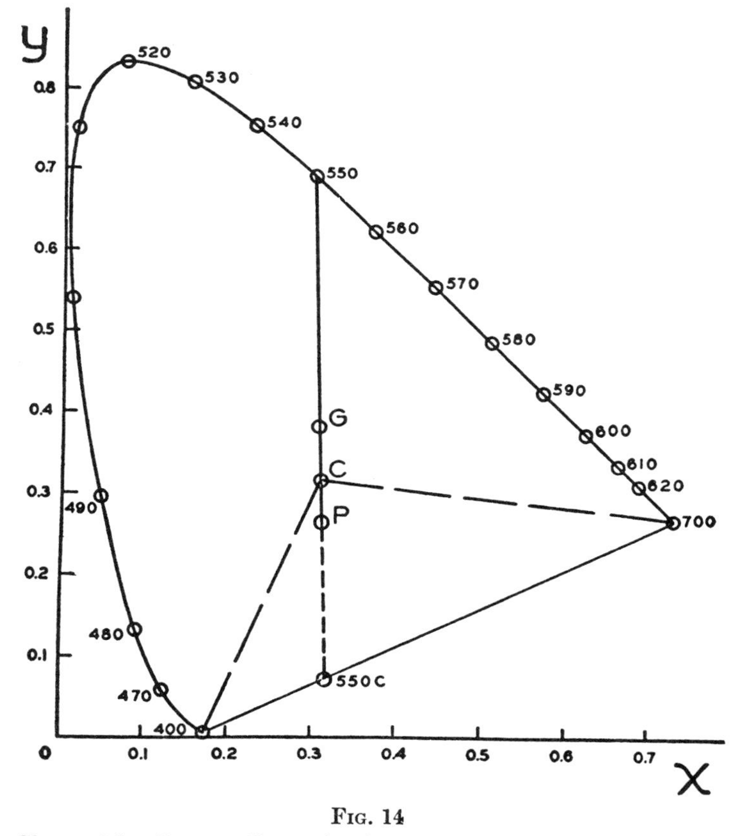

Fig. 14

Chromaticity diagram illustrating how the dominant wavelength of the purple P is specified in terms of that of its complementary color G.

will later be demonstrated, that the complement of a color always lies on the opposite side of the illuminant point on the line through the illuminant point. Thus, in Fig. 14 the complement of the purple P is a green G whose dominant wavelength is 550 millimicrons. The dominant wavelength assigned to this purple is generally written 550c to distinguish the purple from its complementary.

The concept of purity is likewise inapplicable to a purple unless some arbitrary procedure is adopted. The most rational procedure is to regard the lower boundary of the area of real colors as the locus of purples of highest purity. Assigning a purity of 100% to any color lying on this line, the purity of the color represented at P is found to be 21%.

A color specification in terms of brightness, dominant wavelength, and purity[16] can be readily understood because these parameters correspond so closely to the three psychological attributes — brilliance (value), hue, and saturation (chroma). There can be no exclusively psychological system of color specification because, in the end, such systems must be based on physical standards. It is futile to attempt to construct a fundamental color language on the basis of samples of various colors because of the questionable permanence of the samples. Nevertheless, it is interesting to note that the attempts to arrange collections of colored objects in an orderly manner have led to the identification of three attributes which are strikingly similar to brightness, dominant wavelength, and purity.

Table II below may be of some assistance to those who are accustomed to associate color names with certain colors. In this table are listed the brightness, dominant wavelength, and purity of a few colors that have been used so widely that some degree of standardization of nomenclature has been achieved. It must be emphasized that this table is not intended for use as a dictionary to translate one color language into another; neither is the table to be interpreted as an attempt to standardize color nomenclature. It is included merely as an illustration of the reasonableness of this type of color specification.

TABLE II

	Brightness	*Dominant Wavelength*	*Purity*
Cardinal	9%	617 $m\mu$	55%
Baby Pink	50	610	10
Old Rose	23	608	20
Blood Red	17	600	65
Maroon	7	600	35
Henna	8	587	60
Chocolate	5	586	30
Russet	12	585	45
Goldenrod	45	576	70
Ivory	55	575	25
Olive Green	14	572	45
Apple Green	35	568	40
Turquoise	32	500	30
Lupine	32	476	27
Slate	12	476	10
Navy Blue	3	475	20
Royal Purple	4	560c	50
Orchid	33	500c	16

As a matter of interest, the colors whose spectral reflection curves were reproduced in Figs. 1 to 6 have been evaluated in terms of brightness, dominate wavelength, and purity in Table III.

TABLE III

Figure No.	*Brightness*	*Dominant Wavelength*	*Purity*
1	24%	506 $m\mu$	20%
2	33	483	31
3	63	576	80
4	32	595	87
5	13	619	60
6	46	550c	21

11. Color Tolerances

The value of a fundamental color language expressible in numerical terms is immediately obvious, but it is not always appreciated that such a language provides the solution to another important problem — namely, the specification of color tolerances. In every field where a numerical system of mensuration has been achieved, the custom of specifying tolerances has soon been established. For example, no one conversant with machine shop practice would submit a drawing indicating a set of dimensions without at the same time indicating the allowable departure from these nominal dimensions. The

[16] The procedure of regarding every color as a mixture of Illuminant C and spectrum light of some wavelength is the basis of the monochromatic method of colorimetry. Colorimeters have been constructed in which one half of a photometric field is filled with the unknown color and the other half with a mixture of controlled amounts of "white light" and controlled amounts of spectrum light of adjustable wavelength. By adjusting the controls, the color of the two halves of the field can be made to match. In the case of purple, the match is made by adding the spectral component (usually green) to the sample side of the field. In other words, purple can be regarded as a mixture of white light and a negative amount of green light of the proper wavelength. In any case, the dominant wavelength of the unknown color is determined directly from the wavelength scale of the instrument. The brightness of the unknown color is the sum (or difference) of the brightness of the white and the brightness of the spectral component. The purity is the ratio of the brightness of the spectral component to the sum (or difference) of the brightness of the white and the brightness of the spectral component.

The purity determined in this manner is called the colorimetric purity. Because the values of colorimetric purity are negative in the case of purple samples and the lines of constant purity on a chromaticity diagram are discontinuous, the previous definition of purity is to be preferred. For want of a better term, the purity defined in terms of the position on the chromaticity diagram has been called the excitation purity. Throughout this volume, the term *purity* will mean *excitation purity*. The use of excitation purity rather than colorimetric purity is a hardship only on those who use monochromatic colorimeters. For reasons similar to those pertaining to colorimeters in general, monochromatic colorimeters are rarely used.

machinist, in the absence of any information concerning the tolerances, would not know whether the part was intended to be used as a Johansson gauge, which can deviate by less than 0.00001 of an inch from its nominal value, or whether the crudest sort of machine work would be satisfactory. In the field of color, the use of the expression "color match" without an indication of the allowable departure is equally absurd and is the cause of a great deal of misunderstanding. Although some work has already been done in the direction of establishing color tolerances which the eye can just perceive under the most favorable conditions, the results are not included here for the reason that they have not been subjected to any attempt at universal standardization. Also, to include these data might be misleading because there are many applications of colored materials where a tolerance many times larger than the minimum perceptible color difference would be satisfactory. It may be remarked in passing that a method of specifying color tolerances is especially useful as a means of indicating the extent to which a color can be allowed to vary with time as a result of exposure to light or to some other deleterious agent.

The subject matter of this chapter has been presented in some detail in order that the reader who is approaching this subject for the first time may be assured of the rigor of every step of the procedure. Some readers may be disappointed to find that this approach is based entirely upon the use of a spectrophotometer as the fundamental measuring instrument. Such an approach is inevitable because the eye is not an analytical instrument. A specification of color in terms of equivalent stimuli may at times be useful, but the vast majority of color problems require ultimately the wavelength by wavelength analysis furnished by a spectrophotometer. For the benefit of those who lack spectrophotometric facilities, it may be stated that the concepts associated with spectrophotometry are of considerable usefulness in themselves. Also, there is enough current interest in this subject to justify a belief that spectrophotometers will become more readily available. This will result partly from improvements in the design of the instrument, which will effect economies in its production without sacrifice of precision, and partly from an increase in the number of testing laboratories that offer this type of service.

CHAPTER II

SOURCES OF LIGHT

Every source of light is merely a group of radiating atoms. If one atom could be isolated for study, it would be found to emit radiation of a single frequency or wavelength during any interval throughout which it is radiating. This type of radiation is often called *homogeneous* or *monochromatic*.[1] An atom is capable of radiating energy only when it has previously absorbed energy as a result of thermal, electrical, or some other form of excitation. Extensive experiments have shown that an atom is capable of absorbing only certain amounts of energy, each amount being associated with a definite state of the electrons of which the atom is composed. If as a result of some form of excitation, an atom finds itself at an energy level higher than the minimum, it is then capable of radiating energy during its transition to a lower level. The frequency or wavelength of this radiation is determined by the difference between the initial and final energy levels. Thus, although an atom can radiate at only a single frequency during a change from one energy level to another, it may later radiate at a different frequency corresponding to a different pair of energy levels. Hence, a group of atoms isolated from one another as they are in the gaseous state will be observed to emit a group of frequencies or wavelengths which, when analyzed by a prism or diffraction grating, produce a spectrum consisting of certain bright lines with darkness between. The wavelengths of these bright lines are characteristic of the nature of the atom; in other words, they are characteristic of the chemical element of which the atom is a unit.

[1] The terms *homogeneous* and *monochromatic* have undesirable connotations. Nevertheless, they are widely used in treatises on optics. From the standpoint of colorimetrics, light of substantially a single frequency or wavelength is better described by the simple expression *spectrum light*. This expression is employed in this volume except where common usage dictates to the contrary.

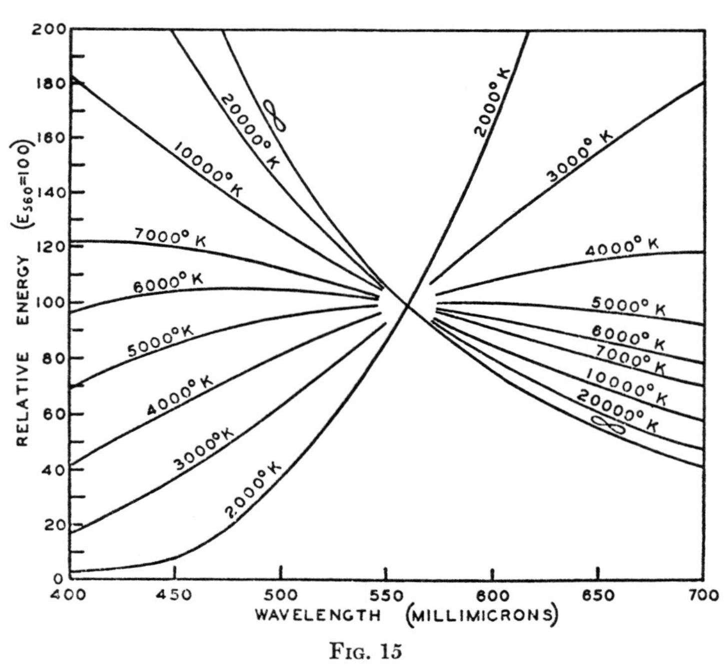

Fig. 15

Relative distribution of energy in the radiation from a black body at various temperatures.

With rare exceptions, the light sources by which colors are observed do not produce line spectra. The common sources, such as the sun, the filament of a tungsten lamp, or the crater of a carbon arc, are incandescent solids. In a solid, the atoms are so closely packed that they are incapable of radiating in their characteristic manner. For this reason, the radiation from an incandescent solid is found to be practically independent of the material of which it is composed, and to depend only on its temperature. The most efficient thermal radiator is a black body, and the closest approach to a completely black body that can be realized experimentally is a small opening in a large cavity whose walls are maintained at a uniform temperature. When the radiation from a black body is analyzed by a prism, it is found that all frequencies or wavelengths are present. A spectrum of this type is called a continuous spectrum.

12. Thermal Radiators

The spectral energy distribution of the radiation from a black body varies with temperature in accordance with Planck's law. This law can be written in the form

$$E = \frac{C_1 \lambda^{-5}}{\epsilon^{\frac{C_2}{\lambda T}} - 1} \qquad (2)$$

where E is the amount of energy radiated in ergs per second per square centimeter of surface within a spectral band one millimicron in width, C_1 is a constant whose value is 3.703×10^{23}, ϵ is the base of the natural system of logarithms, C_2 is a constant whose value is 1.433×10^7, λ is the wavelength of light in millimicrons, and T is the absolute or Kelvin temperature of the body (Centigrade degrees plus 273°). This law is found to hold to the limit of accuracy obtained with the most refined experimental technique. Calculations involving Planck's law are greatly facilitated by use of the data in the International Critical Tables, Volume V, page 237 *et seq.* and also by the use of tables and charts published by the National Bureau of Standards as Miscellaneous Publication No. 56.

The relative distribution of energy in the radiation from black bodies at various temperatures is given in Fig. 15. In conformity with the usual convention, each curve has been plotted on such a scale that the energy is made arbitrarily equal to 100 at 560 millimicrons. Actually the amount of energy radiated within a wavelength band one millimicron wide at 560 millimicrons varies in accordance with the values in Table IV, which has been computed from Planck's law.

The energy radiated by many incandescent solids can be expressed by the equation

$$E = \frac{eC_1\lambda^{-5}}{\epsilon^{\frac{C_2}{\lambda T}} - 1} \qquad (3)$$

which is identical with Planck's law except for the insertion of a constant e, which is known as the emissivity. A tungsten filament, for example, has an emissivity of approximately 0.3 at the ordinary operating temperature of tungsten lamps. This means that a tungsten filament radiates only about ⅓ as much energy at every wavelength as a black body at the same temperature.

TABLE IV

Energy Radiated by a Black Body in the Interval from 559.5 $m\mu$ to 560.5 $m\mu$

Temperature (Degrees Kelvin)	*Energy Radiated (Ergs per second per cm²)*
2000	1.86×10^4
2200	5.98×10^4
2400	1.58×10^5
2600	3.58×10^5
2800	7.22×10^5
3000	1.33×10^6
4000	1.12×10^7
5000	4.05×10^7
6000	9.58×10^7
7000	1.78×10^8
8000	2.86×10^8
9000	4.16×10^8
10,000	5.64×10^8
15,000	1.49×10^9
20,000	2.59×10^9
25,000	3.77×10^9

This proportional reduction in the emission affects only the quantity of radiation and not the quality. Materials whose spectral distribution of energy can be represented by the above equation are known as gray-body radiators.

The chromaticity of the radiation from a black (or gray) body at various temperatures is shown in Fig. 16. It is possible to obtain radiation which is not describable by Planck's law that will nevertheless have the same chromaticity as a black body at some temperature. This temperature is called the color temperature of the radiation. In Fig. 16, the chromaticities of the I.C.I. Illuminants A, B, and C are indicated, as is also the chromaticity of a source E radiating an equal amount of energy at every wavelength. With the exception of Illuminant A, these sources are not identical in chromaticity with a black body at any temperature. They lie so close to the locus of black-body radiators, however, that their chromaticities are indicated with some degree of precision by stating the temperature of the black body to which they most nearly correspond. In the case of the equal-energy source E, the lack of correspondence is so great that the concept of color temperature has little significance. The color temperature of a black- or gray-body radiator is the same as its true (absolute) temperature. For other radiators, color temperature has no connection with the true temperature. The sky, for example, has a color temperature in the neighborhood of 20,000° K.

13. Spectroradiometry

The experimental procedure for determining the energy distribution of a source of light is known as spectroradiometry. In principle, the method consists in dispersing the radiation into its spectral components, allowing the radiation within a narrow wavelength interval to fall on a thermopile or similar device, and noting the temperature rise. The thermopile is then moved to a different position in the spectrum, and the operation is repeated until the entire spectrum has been examined. Although the principles involved in spectroradiometric measurements are extremely simple, the amount of energy available is ordinarily so small that the technique of making accurate measurements can be acquired only by long experience.

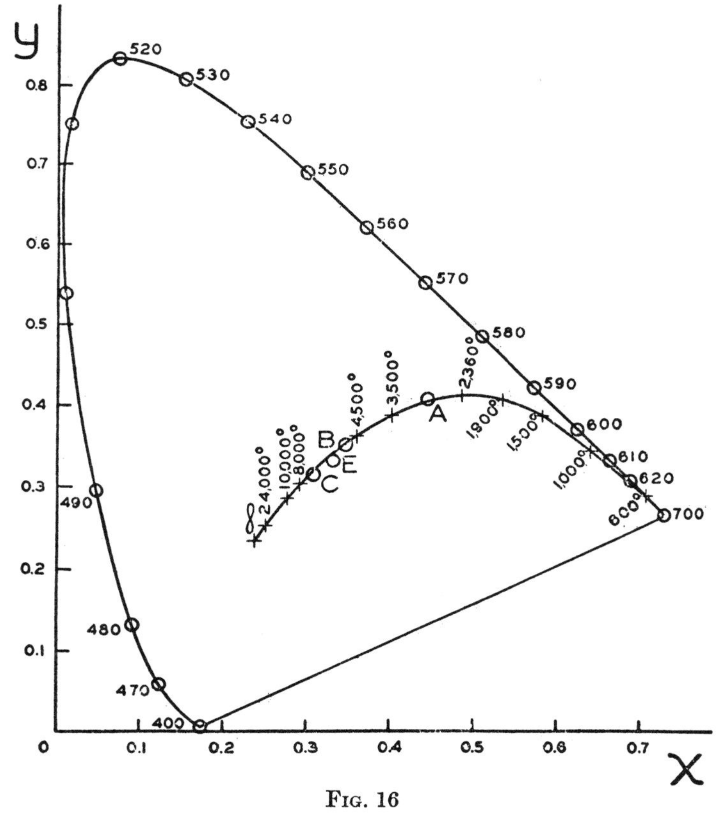

Fig. 16

Diagram showing the chromaticities of various important illuminants. The solid line is the locus of radiation from a black body at various temperatures. The points A, B, and C represent the chromaticity of the I.C.I. Illuminants. The point E represents the chromaticity of a source radiating equal amounts of energy in each wavelength interval.

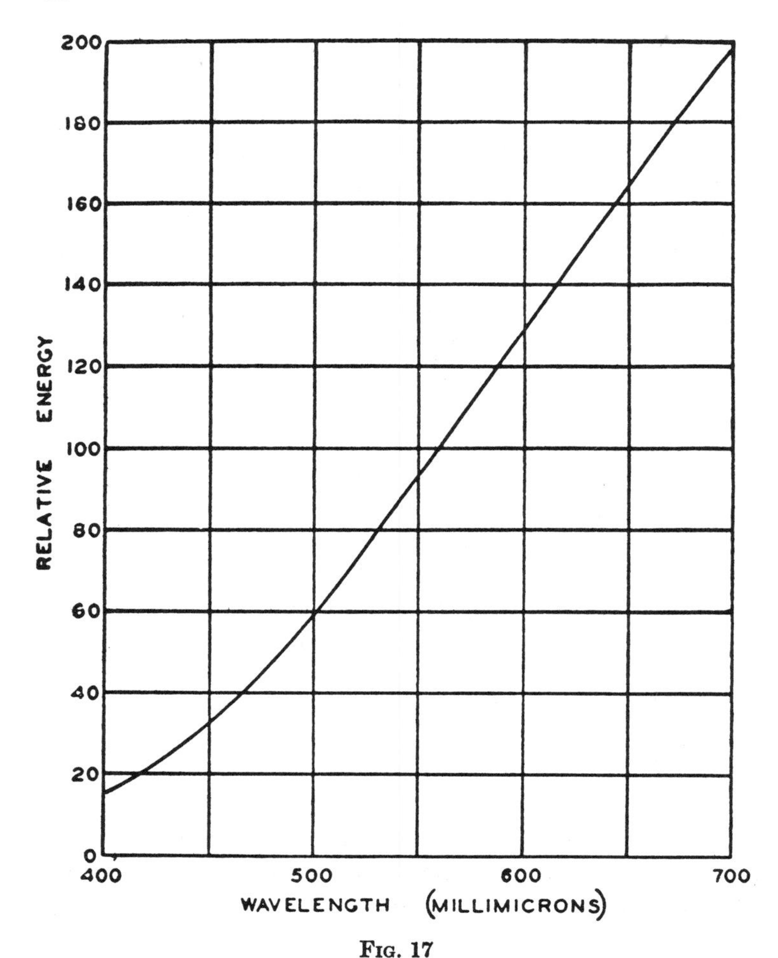

Fig. 17

Relative distribution of energy in the radiation from I.C.I. Illuminant A, which is a gray-body radiator having a temperature of 2848° K.

In fact, there is probably no other type of measurement that contains so many pitfalls for the novice; and even the simplest measurements should not be undertaken without a thorough acquaintance with the literature of the subject.

14. Illuminant A

Because spectroradiometric measurements are so difficult, the International Commission on Illumination selected for use in colorimetry three light sources that can be easily reproduced. These sources are known as I.C.I. Illuminants A, B, and C. Illuminant A is a tungsten lamp operated at a temperature of 2848° K. Suitable lamps that have been properly aged and calibrated are obtainable from the National Bureau of Standards and elsewhere. The data furnished with each lamp enable the user to operate it at the proper temperature without installing the equipment required for color temperature determinations. The relative distribution of energy in Illuminant A is given in column 2 of Table V for every millimicron. The original I.C.I. recommendations listed these values at 5 millimicron intervals only. In preparing Table V, the I.C.I. values were interpolated to one-millimicron intervals by the third difference osculatory formulae.[2] A curve showing the spectral distribution of energy in the radiation from Illuminant A is shown in Fig. 17.

It will be noticed that the last three figures in Table V are separated from the others by a space. The I.C.I. data were given to four significant figures and no additional precision is achieved by the interpolation process. However, since this table is the basis on which other tables are constructed, it is desirable to avoid rejection errors by preserving the additional figures. For ordinary purposes, the last three figures should be ignored.

15. Illuminant B

Illuminant B utilizes a lamp having the spectral quality of Illuminant A in combination with a filter. The filter consists of a layer one centimeter thick of each of the solutions, B_1 and B_2, these solutions being contained in a double cell constructed of white optical glass. The composition of each solution is as follows:

Solution B_1.

Copper sulphate ($CuSO_4.5H_2O$).....	2.452 g.
Mannite ($C_6H_8(OH)_6$)............	2.452 g.
Pyridine (C_5H_5N)................	30.0 c.c.
Distilled water to make...........	1000.0 c.c.

Solution B_2.

Cobalt—ammonium sulphate ($CoSO_4.(NH_4)_2SO_4.6H_2O$)........	21.71 g.
Copper sulphate ($CuSO_4.5H_2O$).....	16.11 g.
Sulphuric acid (density 1.835)......	10.0 c.c.
Distilled water to make...........	1000.0 c.c.

The energy distribution of Illuminant B is given in column 3 of Table V at wavelength intervals of one millimicron. These values were obtained by an interpolation process identical with that used in computing the values for Illuminant A.

16. Illuminant C

Illuminant C likewise consists of a source having the spectral quality of Illuminant A in combination with a filter. An identical cell is used, but the solutions, C_1 and C_2, have in this case the following compositions:

Solution C_1.

Copper sulphate ($CuSO_4.5H_2O$).....	3.412 g.
Mannite ($C_6H_8(OH)_6$)............	3.412 g.
Pyridine (C_5H_5N)................	30.0 c.c.
Distilled water to make...........	1000.0 c.c.

Solution C_2.

Cobalt—ammonium sulphate. ($CoSO_4.(NH_4)_2SO_4.6H_2O$)........	30.580 g.
Copper sulphate ($CuSO_4.5H_2O$).....	22.520 g.
Sulphuric acid (density 1.835).....	10.0 c.c.
Distilled water to make...........	1000.0 c.c.

[2] A detailed description of the use of these formulae may be found in "Extension of the Standard Visibility Data to Intervals of 1 Millimicron by Third Difference Osculatory Interpolation," by Deane B. Judd, *Bureau of Standards Journal of Research.* V. 6, p. 465 (1931).

The distribution of energy from this source is given in column 4 of Table V at wavelength intervals of one millimicron. These values have likewise been obtained by the third difference osculatory interpolation formulae. A curve showing the relative energy of Illuminant C is given in Fig. 7 of Chapter I.

17. Sources of Light for Color Matching

Everyone who has had any occasion to match colors recognizes that an artificial source of constant intensity and quality is preferable to daylight, which varies considerably in both respects and is not always available. Due to a common misconception, the demand is frequently made that the artificial source be so constructed that, when colors appear to match under its light, they will also match under the light from all other sources. No source can ever be found that will satisfy the latter requirement. If a match must be valid under all types of illumination, the only infallible procedure is to examine the match successively under spectrum light of all wavelengths. This, in effect, is the procedure used in spectrophotometry.

It is frequently sufficient to require that the match be valid for a certain temperature range of black-body radiation. In this case, examining the match successively under two properly chosen sources is ordinarily adequate. One source should preferably correspond to a color temperature only slightly higher than that of the yellowest light source that would normally be encountered; the other should correspond to a color temperature only slightly lower than that of the bluest light source that would normally be encountered. If the match is valid under both sources, there is a strong probability that it will be valid under the light from a thermal radiator at any intermediate temperature. To set the limits over which color matches must ordinarily be valid, it is necessary to have information concerning the various aspects of both artificial light and natural light.

18. Tungsten Lamps

Small tungsten lamps of the vacuum type when operated at their rated voltage have a color temperature of about 2400° K. The larger lamps of the gas-filled type have a normal operating color temperature which varies from about 2670° K in the smaller sizes to slightly more than 3000° K in the larger sizes. These values apply to the bare lamp, and the spectral distribution of the radiation illuminating a material under examination may be considerably altered by shades and reflectors or even by reflection from the walls of the room. For most practical purposes, however, Illuminant A may be considered to be a fair representation of the spectral quality of tungsten lamps.

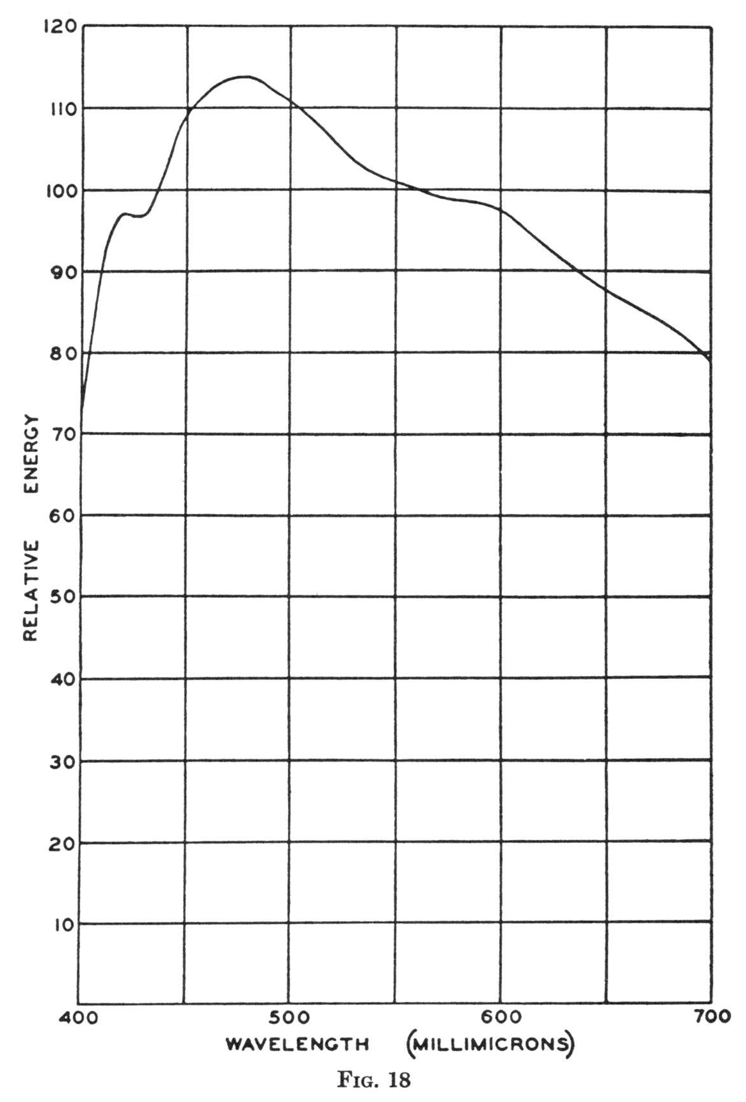

Fig. 18

Relative distribution of energy in sunlight above the earth's atmosphere. (See Table VI.)

19. Solar Radiation

Above the earth's atmosphere, the radiation from the sun is substantially constant in both quality and quantity. Any measurements made on the energy distribution of sunlight must be made under the atmosphere but, by making measurements with the sun at several different elevations, it is possible to compute the atmospheric correction. Fig. 18 shows the relative spectral distribution of energy in sunlight after correcting for the effect of the atmosphere. The values from which the curve in this figure was plotted are given in Table VI.

On a clear day, solar radiation is scattered selectively in its passage through the earth's atmosphere. This selective scattering causes the sky to be blue, and causes the radiations of short wavelengths to be correspondingly reduced in intensity in the unscattered beam.[3] The spectral distribution of energy in mean noon sunlight at Washington is given in Table VII and is represented

[3] At noon on a clear summer day, the illumination on a horizontal surface due to direct sunlight is approximately 8000 foot-candles and that due to the diffuse skylight about 2000 foot-candles.

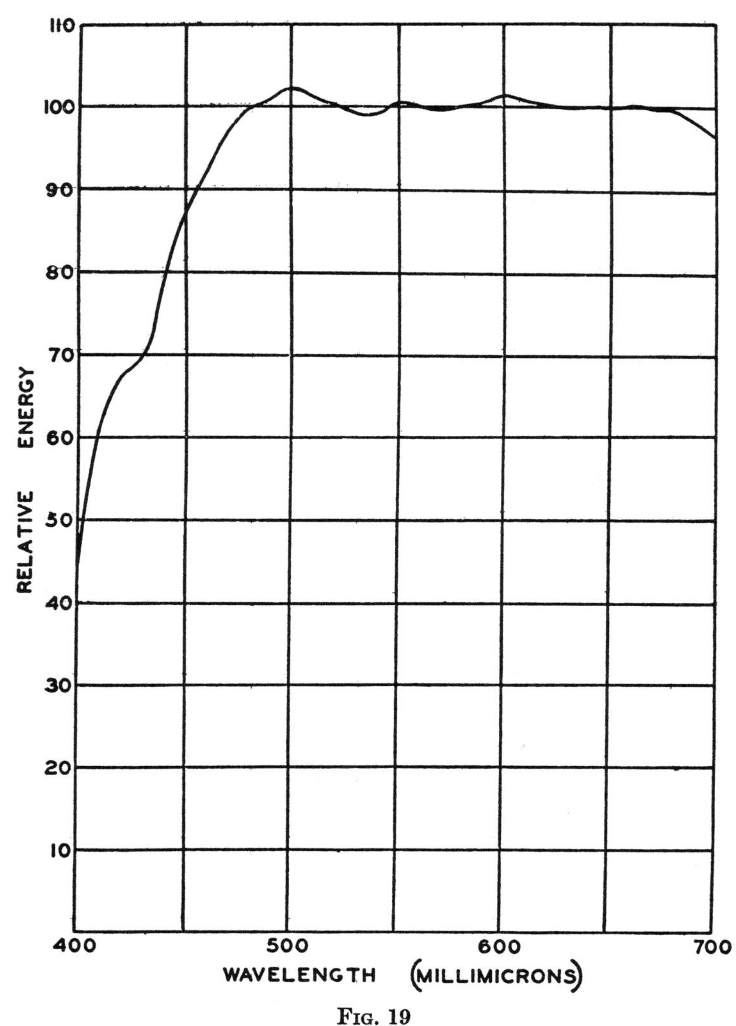

FIG. 19
Relative distribution of energy in mean noon sunlight at Washington. (See Table VII.)

graphically in Fig. 19. In the limit, if the scattering particles in the atmosphere are very small, the scattering follows Rayleigh's inverse fourth-power law. This means that radiation with a wavelength of 400 millimicrons has a scattering coefficient nearly ten times as great as radiation with a wavelength of 700 millimicrons. If solar radiation outside the atmosphere had a strictly uniform distribution of energy throughout the spectrum, the distribution of energy in skylight, if due to Rayleigh scattering, would be

$$E = \frac{a\ constant}{\lambda^4}. \qquad (4)$$

By comparing this equation with Planck's law (eq. 2), it can be shown that such radiation is similar in quality to the radiation that a black body would emit if it could be raised to an infinite temperature.

It rarely happens that the conditions for Rayleigh scattering are fulfilled by the atmosphere. In the extreme case of a completely overcast day, the scattering becomes almost non-selective as to wavelength. The radiation from any part of the sky is then a close approximation in spectral quality to solar radiation outside the earth's atmosphere. Light from the north sky, which is frequently used as a source for color matching, is therefore subject to large variations in quality with changing weather conditions, the limits corresponding to a black body at a nearly infinite temperature on the one hand, and to a black body at about 6500° K on the other.

By definition, daylight is the light from the entire sky including direct sunlight if the sky is clear. The spectral quality of daylight is a close approximation to solar radiation outside the atmosphere, regardless of the state of the weather or the altitude of the sun. For this reason, Illuminant C, which rather closely approximates daylight, is the preferred source on which to base the color language discussed in the preceding chapter.

TABLE V

Energy Distribution of Illuminants A, B, and C

Wavelength	E_A	E_B	E_C	*Wavelength*	E_A	E_B	E_C
380 mμ	9.79 000	22.40 000	33.00 000	435 mμ	26.64 000	77.31 000	117.75 000
81	9.98 000	23.22 000	34.32 000	36	27.04 480	78.06 768	118.63 120
82	10.21 000	24.09 500	35.69 000	37	27.45 320	78.79 864	119.45 160
83	10.43 800	24.98 500	37.08 000	38	27.86 520	79.49 976	120.20 640
84	10.66 500	25.96 000	38.45 000	39	28.28 080	80.16 792	120.89 080
385	10.90 000	26.85 000	39.92 000	440	28.70 000	80.80 000	121.50 000
86	11.13 160	27.73 312	41.37 552	41	29.12 280	81.39 328	122.02 760
87	11.36 640	28.60 936	42.85 776	42	29.54 920	81.94 984	122.47 680
88	11.60 440	29.48 904	44.36 024	43	29.97 920	82.47 376	122.85 720
89	11.84 560	30.38 248	45.87 648	44	30.41 280	82.96 912	123.17 840
390	12.09 000	31.30 000	47.40 000	445	30.85 000	83.44 000	123.45 000
91	12.33 760	32.24 464	48.92 968	46	31.29 096	83.88 320	123.66 480
92	12.58 840	33.20 952	50.46 984	47	31.73 568	84.29 600	123.81 640
93	12.84 240	34.19 008	52.02 216	48	32.18 392	84.68 320	123.91 560
94	13.09 960	35.18 176	53.58 832	49	32.63 544	85.04 960	123.97 320
395	13.36 000	36.18 000	55.17 000	450	33.09 000	85.40 000	124.00 000
96	13.62 344	37.18 544	56.76 688	51	33.54 744	85.72 768	123.98 240
97	13.88 992	38.20 112	58.37 784	52	34.00 792	86.02 944	123.91 320
98	14.15 968	39.22 608	60.00 336	53	34.47 168	86.31 536	123.81 280
99	14.43 296	40.25 936	61.64 392	54	34.93 896	86.59 552	123.70 160
400	14.71 000	41.30 000	63.30 000	455	35.41 000	86.88 000	123.60 000
1	14.99 080	42.34 864	64.97 320	56	35.88 512	87.16 208	123.49 520
2	15.27 520	43.40 592	66.66 320	57	36.36 416	87.43 504	123.37 360
3	15.56 320	44.47 088	68.36 760	58	36.84 664	87.70 896	123.25 440
4	15.85 480	45.54 256	70.08 400	59	37.33 208	87.99 392	123.15 680
405	16.15 000	46.62 000	71.81 000	460	37.82 000	88.30 000	123.10 000
6	16.44 896	47.70 384	73.54 784	61	38.31 008	88.63 072	123.09 040
7	16.75 168	48.79 472	75.29 912	62	38.80 264	88.97 936	123.11 520
8	17.05 792	49.89 168	77.06 048	63	39.29 816	89.34 064	123.16 480
9	17.36 744	50.99 376	78.82 856	64	39.79 712	89.70 928	123.22 960
410	17.68 000	52.10 000	80.60 000	465	40.30 000	90.08 000	123.30 000
11	17.99 528	53.21 392	82.38 280	66	40.80 696	90.45 776	123.38 416
12	18.31 344	54.33 616	84.17 920	67	41.31 768	90.84 608	123.48 848
13	18.63 496	55.46 144	85.97 720	68	41.83 192	91.23 752	123.60 072
14	18.96 032	56.58 448	87.76 480	69	42.34 944	91.62 464	123.70 864
415	19.29 000	57.70 000	89.53 000	470	42.87 000	92.00 000	123.80 000
16	19.62 432	58.81 168	91.28 096	71	43.39 360	92.36 568	123.87 912
17	19.96 296	59.92 304	93.02 568	72	43.92 040	92.72 664	123.95 416
18	20.30 544	61.02 856	94.75 192	73	44.45 040	93.07 976	124.01 864
19	20.65 128	62.12 272	96.44 744	74	44.98 360	93.42 192	124.06 608
420	21.00 000	63.20 000	98.10 000	475	45.52 000	93.75 000	124.09 000
21	21.35 144	64.26 216	99.71 328	76	46.05 992	94.06 592	124.09 536
22	21.70 592	65.31 288	101.29 544	77	46.60 336	94.37 176	124.08 648
23	22.06 368	66.34 952	102.84 096	78	47.14 984	94.66 464	124.05 592
24	22.42 496	67.36 944	104.34 432	79	47.69 888	94.94 168	123.99 624
425	22.79 000	68.37 000	105.80 000	480	48.25 000	95.20 000	123.90 000
26	23.15 880	69.35 248	107.21 040	81	48.80 288	95.44 504	123.77 440
27	23.53 120	70.31 864	108.57 920	82	49.35 784	95.67 872	123.62 440
28	23.90 720	71.26 656	109.90 280	83	49.91 536	95.89 288	123.43 920
29	24.28 680	72.19 432	111.17 760	84	50.47 592	96.07 936	123.20 800
430	24.67 000	73.10 000	112.40 000	485	51.04 000	96.23 000	122.92 000
31	25.05 680	73.98 680	113.57 560	86	51.60 776	96.34 960	122.58 064
32	25.44 720	74.85 600	114.70 680	87	52.17 888	96.44 360	122.19 712
33	25.84 120	75.70 280	115.78 520	88	52.75 312	96.50 480	121.76 128
34	26.23 880	76.52 240	116.80 240	89	53.33 024	96.52 600	121.26 496

TABLE V (*Continued*)

Wavelength	E_A	E_B	E_C	*Wavelength*	E_A	E_B	E_C
490 $m\mu$	53.91 000	96.50 000	120.70 000	550 $m\mu$	92.91 000	101.00 000	105.20 000
91	54.49 240	96.42 136	120.05 712	51	93.61 360	101.29 056	105.35 736
92	55.07 760	96.29 488	119.34 176	52	94.31 840	101.55 488	105.48 448
93	55.66 560	96.12 872	118.56 784	53	95.02 440	101.79 392	105.57 992
94	56.25 640	95.93 104	117.74 928	54	95.73 160	102.00 864	105.64 224
495	56.85 000	95.71 000	116.90 000	555	96.44 000	102.20 000	105.67 000
96	57.44 656	95.45 920	116.00 912	56	97.14 976	102.36 608	105.66 288
97	58.04 608	95.17 320	115.06 736	57	97.86 088	102.50 624	105.62 184
98	58.64 832	94.86 160	114.09 104	58	98.57 312	102.62 336	105.54 736
99	59.25 304	94.53 400	113.09 648	59	99.28 624	102.72 032	105.43 992
500	59.86 000	94.20 000	112.10 000	560	100.00 000	102.80 000	105.30 000
1	60.46 920	93.85 192	111.08 944	61	100.71 440	102.86 176	105.12 440
2	61.08 080	93.48 336	110.05 392	62	101.42 960	102.90 368	104.91 280
3	61.69 480	93.10 584	109.01 168	63	102.14 560	102.92 672	104.67 000
4	62.31 120	92.73 088	107.98 096	64	102.86 240	102.93 184	104.40 080
505	62.93 000	92.37 000	106.98 000	565	103.58 000	102.92 000	104.11 000
6	63.55 120	92.01 584	105.99 680	66	104.29 840	102.89 024	103.79 312
7	64.17 480	91.66 072	105.01 920	67	105.01 760	102.84 192	103.44 696
8	64.80 080	91.31 568	104.06 520	68	105.73 760	102.77 648	103.07 824
9	65.42 920	90.99 176	103.15 280	69	106.45 840	102.69 536	102.69 368
510	66.06 000	90.70 000	102.30 000	570	107.18 000	102.60 000	102.30 000
11	66.69 320	90.43 592	101.50 056	71	107.90 240	102.48 752	101.89 496
12	67.32 880	90.19 216	100.74 248	72	108.62 560	102.35 696	101.47 408
13	67.96 680	89.97 544	100.03 512	73	109.34 960	102.21 264	101.04 072
14	68.60 720	89.79 248	99.38 784	74	110.07 440	102.05 888	100.59 824
515	69.25 000	89.65 000	98.81 000	575	110.80 000	101.90 000	100.15 000
16	69.89 552	89.54 512	98.29 824	76	111.52 672	101.73 328	99.69 312
17	70.54 376	89.47 336	97.84 632	77	112.25 456	101.55 584	99.22 536
18	71.19 424	89.43 904	97.45 928	78	112.98 304	101.37 176	98.75 104
19	71.84 648	89.44 648	97.14 216	79	113.71 168	101.18 512	98.27 448
520	72.50 000	89.50 000	96.90 000	580	114.44 000	101.00 000	97.80 000
21	73.15 464	89.60 344	96.74 000	81	115.16 784	100.81 496	97.32 504
22	73.81 072	89.75 392	96.65 880	82	115.89 552	100.62 728	96.84 672
23	74.46 848	89.94 568	96.64 560	83	116.62 328	100.43 912	96.36 888
24	75.12 816	90.17 296	96.68 960	84	117.35 136	100.25 264	95.89 536
525	75.79 000	90.43 000	96.78 000	585	118.08 000	100.07 000	95.43 000
26	76.45 400	90.72 240	96.92 672	86	118.80 920	99.89 040	94.97 104
27	77.12 000	91.05 400	97.13 696	87	119.53 880	99.71 240	94.51 592
28	77.78 800	91.41 640	97.39 584	88	120.26 880	99.53 720	94.06 728
29	78.45 800	91.80 120	97.68 848	89	120.99 920	99.36 600	93.62 776
530	79.13 000	92.20 000	98.00 000	590	121.73 000	99.20 000	93.20 000
31	79.80 416	92.61 776	98.33 840	91	122.46 152	99.03 584	92.78 064
32	80.48 048	93.06 008	98.71 360	92	123.19 376	98.87 272	92.36 792
33	81.15 872	93.51 952	99.11 360	93	123.92 624	98.71 568	91.96 688
34	81.83 864	93.98 864	99.52 640	94	124.65 848	98.56 976	91.58 256
535	82.52 000	94.46 000	99.94 000	595	125.39 000	98.44 000	91.22 000
36	83.20 296	94.93 936	100.36 288	96	126.12 048	98.32 320	90.87 616
37	83.88 768	95.43 168	100.80 304	97	126.85 024	98.21 600	90.54 768
38	84.57 392	95.92 832	101.24 776	98	127.57 976	98.12 320	90.23 912
39	85.26 144	96.42 064	101.68 432	99	128.30 952	98.04 960	89.95 504
540	85.95 000	96.90 000	102.10 000	600	129.04 000	98.00 000	89.70 000
41	86.63 944	97.37 024	102.49 944	1	129.77 168	97.97 728	89.47 736
42	87.32 992	97.83 712	102.89 112	2	130.50 424	97.97 824	89.28 408
43	88.02 168	98.29 488	103.26 808	3	131.23 696	97.99 856	89.11 512
44	88.71 496	98.73 776	103.62 336	4	131.96 912	98.03 392	88.96 544
545	89.41 000	99.16 000	103.95 000	605	132.70 000	98.08 000	88.83 000
46	90.10 696	99.56 512	104.25 088	6	133.42 912	98.14 000	88.71 232
47	90.80 568	99.95 696	104.53 064	7	134.15 696	98.21 680	88.61 576
48	91.50 592	100.33 024	104.78 496	8	134.88 424	98.30 560	88.53 504
49	92.20 744	100.67 968	105.00 952	9	135.61 168	98.40 160	88.46 988

TABLE V (*Continued*)

Wavelength	E_A	E_B	E_C	*Wavelength*	E_A	E_B	E_C
610 $m\mu$	136.34 000	98.50 000	88.40 000	670 $m\mu$	178.77 000	104.90 000	86.30 000
11	137.06 968	98.60 176	88.34 200	71	179.44 320	104.84 568	86.10 992
12	137.80 024	98.71 008	88.29 440	72	180.11 480	104.78 664	85.92 016
13	138.53 096	98.82 352	88.25 480	73	180.78 480	104.71 976	85.72 544
14	139.26 112	98.94 064	88.22 080	74	181.45 320	104.64 192	85.52 048
615	139.99 000	99.06 000	88.19 000	675	182.12 000	104.55 000	85.30 000
16	140.71 760	99.18 256	88.16 352	76	182.78 520	104.44 576	85.06 704
17	141.44 440	99.30 928	88.14 296	77	183.44 880	104.33 128	84.82 512
18	142.17 040	99.43 872	88.12 664	78	184.11 080	104.20 392	84.56 968
19	142.89 560	99.56 944	88.11 288	79	184.77 120	104.06 104	84.29 616
620	143.62 000	99.70 000	88.10 000	680	185.43 000	103.90 000	84.00 000
21	144.34 344	99.83 104	88.08 912	81	186.08 720	103.71 712	83.67 688
22	145.06 592	99.96 352	88.08 136	82	186.74 280	103.51 416	83.32 984
23	145.78 768	100.09 648	88.07 504	83	187.39 680	103.29 664	82.96 536
24	146.50 896	100.22 896	88.06 848	84	188.04 920	103.07 008	82.58 992
625	147.23 000	100.36 000	88.06 000	685	188.70 000	102.84 000	82.21 000
26	147.95 080	100.49 056	88.05 056	86	189.34 920	102.60 320	81.82 128
27	148.67 120	100.62 128	88.04 128	87	189.99 680	102.35 600	81.41 944
28	149.39 120	100.75 072	88.03 072	88	190.64 280	102.10 320	81.01 096
29	150.11 080	100.87 744	88.01 744	89	191.28 720	101.84 960	80.60 232
630	150.83 000	101.00 000	88.00 000	690	191.93 000	101.60 000	80.20 000
31	151.54 912	101.11 584	87.97 584	91	192.57 136	101.35 568	79.80 448
32	152.26 816	101.22 592	87.94 592	92	193.21 128	101.11 344	79.41 144
33	152.98 664	101.33 408	87.91 408	93	193.84 952	100.87 136	79.02 016
34	153.70 408	101.44 416	87.88 416	94	194.48 584	100.62 752	78.62 992
635	154.42 000	101.56 000	87.86 000	695	195.12 000	100.38 000	78.24 000
36	155.13 440	101.67 952	87.83 888	96	195.75 184	100.12 976	77.85 072
37	155.84 760	101.80 016	87.81 824	97	196.38 152	99.87 808	77.46 256
38	156.55 960	101.92 504	87.80 216	98	197.00 928	99.62 352	77.07 504
39	157.27 040	102.05 728	87.79 472	99	197.62 536	99.36 464	76.68 768
640	157.98 000	102.20 000	87.80 000	700	198.26 000	99.10 000	76.30 000
41	158.68 808	102.35 656	87.82 168	1	198.88 336	98.82 928	75.91 232
42	159.39 464	102.52 488	87.85 704	2	199.50 528	98.55 344	75.52 496
43	160.10 016	102.69 992	87.90 056	3	200.12 552	98.27 296	75.13 744
44	160.80 512	102.87 664	87.94 672	4	200.74 384	97.98 832	74.74 928
645	161.51 000	103.05 000	87.99 000	705	201.36 000	97.70 000	74.36 000
46	162.21 528	103.22 256	88.03 408	6	201.97 400	97.40 800	73.96 992
47	162.92 064	103.39 768	88.08 264	7	202.58 600	97.11 200	73.57 936
48	163.62 536	103.57 152	88.13 016	8	203.19 600	96.81 200	73.18 784
49	164.32 872	103.74 024	88.17 112	9	203.80 400	96.50 800	72.79 488
650	165.03 000	103.90 000	88.20 000	710	204.41 000	96.20 000	72.40 000
51	165.72 904	104.05 272	88.21 824	11	205.01 400	95.88 800	72.00 416
52	166.42 632	104.20 096	88.22 952	12	205.61 600	95.57 200	71.60 768
53	167.12 208	104.34 184	88.23 168	13	206.21 600	95.25 200	71.20 912
54	167.81 656	104.47 248	88.22 256	14	206.81 400	94.92 800	70.80 704
655	168.51 000	104.59 000	88.20 000	715	207.41 000	94.60 000	70.40 000
56	169.20 240	104.69 520	88.16 528	16	208.00 400	94.26 800	69.98 480
57	169.89 360	104.79 000	88.11 984	17	208.59 600	93.93 200	69.56 240
58	170.58 360	104.87 320	88.06 176	18	209.18 600	93.59 200	69.13 760
59	171.27 240	104.94 360	87.98 912	19	209.77 400	93.24 800	68.71 520
660	171.96 000	105.00 000	87.90 000	720	210.36 000	92.90 000	68.30 000
61	172.64 640	105.04 128	87.79 216	21	210.94 384	92.54 480	67.89 200
62	173.33 160	105.06 824	87.66 688	22	211.52 552	92.18 240	67.48 800
63	174.01 560	105.08 256	87.52 752	23	212.10 528	91.81 760	67.08 800
64	174.69 840	105.08 592	87.37 744	24	212.68 336	91.45 520	66.69 200
665	175.38 000	105.08 000	87.22 000	725	213.26 000	91.10 000	66.30 000
66	176.06 056	105.06 336	87.05 264	26	213.83 552	90.74 880	65.90 880
67	176.74 008	105.03 488	86.87 312	27	214.40 976	90.39 840	65.51 840
68	177.41 832	104.99 672	86.68 528	28	214.98 224	90.05 360	65.13 360
69	178.09 504	104.95 104	86.49 296	29	215.55 248	89.71 920	64.75 920

TABLE V (*Continued*)

Wavelength	E_A	E_B	E_C	*Wavelength*	E_A	E_B	E_C
730 mμ	216.12 000	89.40 000	64.40 000	54 mμ	229.06 816	84.85 600	58.61 600
31	216.68 480	89.09 600	64.05 600	755	229.58 000	84.80 000	58.50 000
32	217.24 720	88.80 400	63.72 400	56	230.09 000	84.75 600	58.39 600
33	217.80 720	88.52 400	63.40 400	57	230.59 800	84.72 400	58.30 400
34	218.36 480	88.25 600	63.09 600	58	231.10 400	84.70 400	58.22 400
735	218.92 000	88.00 000	62.80 000	59	231.60 800	84.69 600	58.15 600
36	219.47 248	87.75 920	62.52 080	760	232.11 000	84.70 000	58.10 000
37	220.02 224	87.53 360	62.25 840	61	232.61 016	84.71 600	58.05 600
38	220.56 976	87.31 840	62.00 560	62	233.10 848	84.74 400	58.02 400
39	221.11 552	87.10 880	61.75 520	63	233.60 472	84.78 400	58.00 400
740	221.66 000	86.90 000	61.50 000	64	234.09 864	84.83 600	57.99 600
41	222.20 352	86.68 880	61.23 520	765	234.59 000	84.90 000	58.00 000
42	222.74 576	86.47 840	60.96 560	66	235.07 880	84.97 760	58.01 920
43	223.28 624	86.27 360	60.69 840	67	235.56 520	85.06 880	58.05 360
44	223.82 448	86.07 920	60.44 080	68	236.04 920	85.17 120	58.09 840
745	224.36 000	85.90 000	60.20 000	69	236.53 080	85.28 240	58.14 880
46	224.89 280	85.73 600	59.97 600	770	237.01 000	85.40 000	58.20 000
47	225.42 320	85.58 400	59.76 400	71	237.45 080	85.52 400	58.24 880
48	225.95 120	85.44 400	59.56 400	72	237.94 920	85.65 600	58.29 840
49	226.47 680	85.31 600	59.37 600	73	238.44 520	85.79 600	58.35 360
750	227.00 000	85.20 000	59.20 000	74	238.93 880	85.94 400	58.41 920
51	227.52 064	85.09 600	59.03 600	75	239.37 000	86.10 000	58.50 000
52	228.03 872	85.00 400	58.88 400	780	241.67 000	87.00 000	59.10 000
53	228.55 448	84.92 400	58.74 400				

TABLE VI [4]

SPECTRAL DISTRIBUTION OF SUNLIGHT ABOVE THE ATMOSPHERE

Wavelength	*Relative Energy*	*Wavelength*	*Relative Energy*
360 mμ	60.0	560mμ	100.0
70	63.8	70	99.1
80	62.0	80	98.6
90	63.9	90	98.3
400	73.4	600	97.4
10	91.5	10	95.2
20	97.0	20	93.1
30	96.9	30	91.0
40	102.9	40	89.3
450	109.6	650	87.5
60	112.0	60	86.0
70	113.5	70	84.6
80	113.6	80	83.3
90	112.1	90	81.4
500	110.7	700	79.1
10	108.5	10	76.8
20	105.9	20	74.4
30	103.4	30	72.2
40	101.7	40	70.2
550	100.9	750	68.2

TABLE VII [4]

SPECTRAL DISTRIBUTION OF MEAN NOON SUNLIGHT AT WASHINGTON

Wavelength	*Relative Energy*	*Wavelength*	*Relative Energy*
360 mμ	17.3	560mμ	100.0
70	21.6	70	99.5
80	24.6	80	100.0
90	29.5	90	100.4
400	44.8	600	101.3
10	60.3	10	100.7
20	67.2	20	100.3
30	69.6	30	99.9
40	78.5	40	100.0
450	86.9	650	99.9
60	92.0	60	100.1
70	97.0	70	99.6
80	99.7	80	99.5
90	100.9	90	98.2
500	102.3	700	96.5
10	101.0	10	94.0
20	100.2	20	92.0
30	99.2	30	89.7
40	99.0	40	88.1
550	100.5	750	85.8

[4] The data in this table are based on measurements by Abbot and associates. The values at ten millimicron intervals were interpolated by K. S. Gibson and associates at the National Bureau of Standards.

CHAPTER III

SPECTROPHOTOMETRY

THE principles underlying spectrophotometric measurements were discussed briefly in Chapter I on the tacit assumption that the test object was opaque. In general, when light falls on an object, a portion of the incident light is reflected, a portion is absorbed within the material, and the remainder is transmitted. Objects are said to be opaque when the transmitted component is so small as to be negligible. The converse case of a transparent object with a negligible reflection factor is never realized. Even a piece of clear white optical glass reflects about 8% of the incident light, absorbs a fraction of 1% in ordinary thicknesses, and transmits nearly 92%. For this reason, an object will be called transparent if the proportion of the incident light that it transmits is too large to be ignored. On this basis, a sheet of paper or a textile fabric must usually be considered transparent.[1]

20. Transmission Factors of Homogeneous Materials

Spectrophotometers were originally devised for the determination of the transmission factors of homogeneous transparent materials. When used for this purpose, the principle underlying their operation consists in first dispersing the light from a suitable source into its spectral components. A nearly monochromatic beam of a known average wavelength is then isolated, and is divided in some suitable manner into two beams of equal intensity. If the test sample is inserted in one of these beams, the intensities will no longer be equal. The balance may be restored, however, by reducing the intensity of the comparison beam through the use of an adjustable diaphragm or an equivalent device. The ratio of the final intensity of the comparison beam to its original intensity can then be read directly from the calibrated scale of the instrument. This ratio is the transmission factor of the sample at this wavelength. The operation is repeated with monochromatic light of other wavelengths in turn until the entire visible spectrum has been adequately examined. A typical transmission curve is shown in Fig. 20.

It is not the purpose of this handbook to discuss the various arrangements of optical parts that have been devised to accomplish the above result. Specific information on this subject can be found in the literature. As a key to the literature, the 1925 Report of the Optical Society Committee on Spectrophotometry[2] will be found invaluable. Also the more recent issues of the *Journal of the Optical Society of America* contain many articles relating to subsequent developments.

In an optical sense, transparent materials may be either homogeneous or non-homogeneous.[3] In measuring the transmission factor of a homogeneous material, such as a piece of colored glass, the usual procedure is to prepare the sample in the form of a flat plate with plane parallel faces. If the material is a liquid, it is placed in a cell with plane parallel walls as illustrated in Fig. 21. Since the amount of absorption depends upon the distance that the light travels through the material, it is important that the length of the optical path be known. Ordinarily the length of the optical path is made equal

[1] No attempt is made here to distinguish between *transparency* and *translucency*. Etymologically, translucency is the better word to use for the generic term. Precedent, however, favors the use of the term transparency. For example, it has long been customary to speak of the transparency of a photographic deposit, even though the scattering of light within such a deposit makes it impossible to obtain a clear view of objects lying behind it. Such distinction as may be necessary between transparency and translucency will be made clear later in connection with the discussion of the methods by which the transmission factor is determined in the case of homogeneous and non-homogeneous materials.

[2] *Journal of the Optical Society of America*, V. 10, p. 169–241 (1925).

[3] In the ordinary use of the terms, a homogeneous material would be called transparent and a non-homogeneous material, translucent.

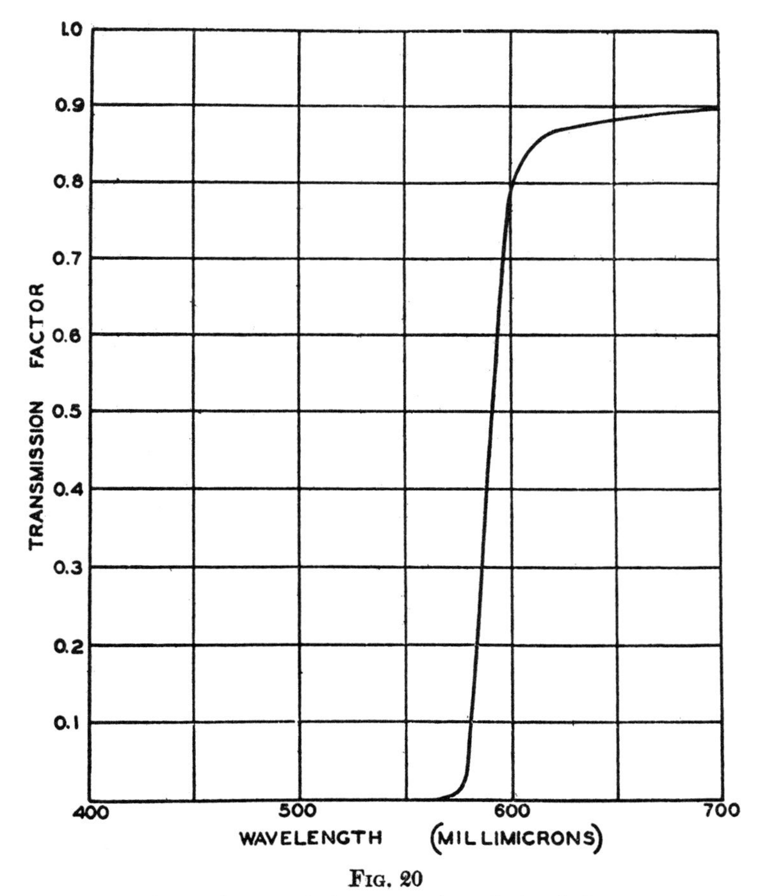

Fig. 20

Spectral transmission curve of a red signal glass.

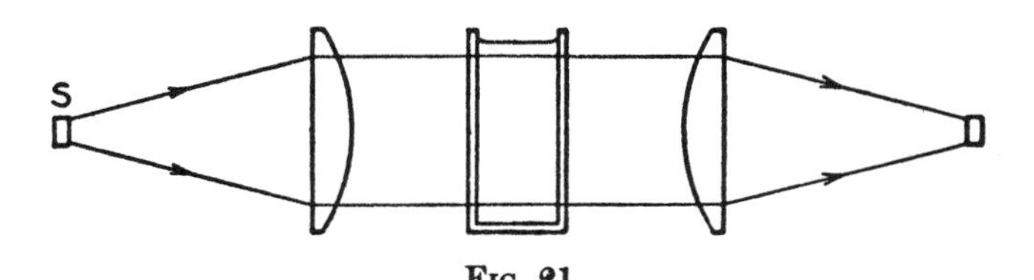

FIG. 21

Schematic diagram of a portion of the optical system employed in measuring the transmission factors of homogeneous materials.

to one dimension of the sample. This is the case in Fig. 21, where a collimated (parallel) beam of light is incident normally on the face of the cell. A truly collimated beam can never be achieved in practice because such a beam could originate only at a source of vanishingly small area. If the source is of finite area, the rays from any point of the source, although parallel to themselves after collimation, are not parallel to the rays that emanate from other points of the source. However, by restricting the beam at S to dimensions that are small relative to the focal lengths of the lenses, the result that would be obtained with a strictly collimated beam may be satisfactorily approximated.

In the arrangement shown in Fig. 21, the absorption by the glass from which the cell walls are composed may not be negligible. Also, every air-glass interface reflects approximately 4% of the light incident upon it. Hence, even if the cell walls were constructed of glass having no absorption and the cell were filled with a clear, non-absorbing liquid, the measured value of the transmission factor would be in the neighborhood of 92%. It is desirable in many applications of spectrophotometry to determine the absorption of a solution having a definite thickness and consisting of a known concentration of a dye in some solvent. To isolate the absorption due to the dye alone, the transmission factor of the cell filled with pure solvent may first be determined. Dividing the measured transmission factor of the cell containing the dye solution by the transmission factor of this "dummy" cell corrects the transmission factor of the dye for losses due to the cell and solvent. An alternative procedure is to secure two identical cells, one of which is filled with the dye solution and the other with the pure solvent. The cell containing the dye solution is placed in the sample beam, and that containing the solvent in the comparison beam of the photometer. The presence of the dummy cell in the comparison beam automatically corrects the values of transmission factor for the absorption and reflection losses in the cell and solvent.

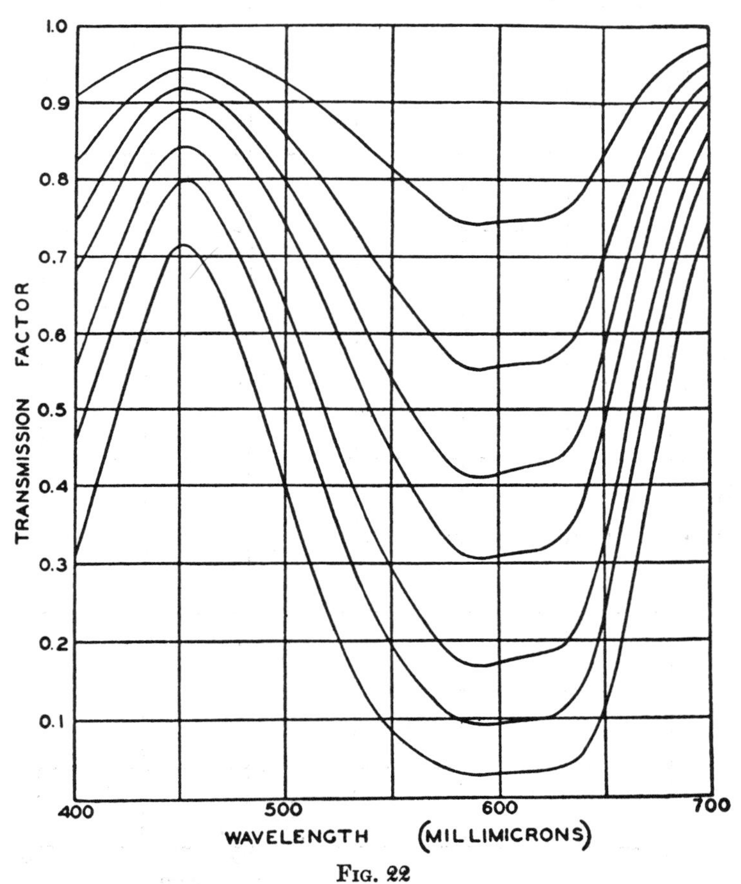

FIG. 22

Spectral transmission curves (corrected for surface reflections) of various thicknesses of a magenta glass. The thicknesses are 1.0 mm., 2.0 mm., 3.0 mm., 4.0 mm., 6.0 mm., 8.0 mm., and 12.0 mm.

Considerations similar to the above apply to the measurement of the transmission factors of solid materials such as colored glasses. The transmission factors that are determined in the manner indicated above for a sample of a given thickness include the reflection losses at the air-glass interfaces. Hence, before the observed transmission factors can be used to calculate the transmission factors for some other thickness of the material by application of Bouguer's law (to be discussed later), a correction must be made for the reflection losses. This correction may be made by measuring two samples of the same material having different thicknesses. From these data the reflection losses can be eliminated by computation. If the index of refraction of a material is known, the reflection losses can generally be computed with adequate precision by application of Fresnel's equations. In the case of normal (perpendicular) incidence, Fresnel's equations reduce to the form

$$R = \left(\frac{n' - n}{n' + n}\right)^2 \qquad (5)$$

where R is the reflection factor of a boundary surface, n is the index of refraction of the medium on one side of the boundary and n' that of the medium on the other side. For the boundary surface between air and a glass with an index of 1.5, the value of R is 0.04. It may be mentioned in passing that, because of inter-surface reflections, the transmission factor of two or more glasses in series is not exactly the product of their several transmission factors.

21. BOUGUER'S LAW

The transmission factor of a homogeneous material, after correction for surface losses, varies with the thickness of the sample in accordance with Bouguer's law.[4] Suppose that a unit thickness of the material has a transmission factor t. A thickness x of the material will then have a transmission factor

$$T = t^x \qquad (6a)$$

This law is frequently written in the form

$$T = \epsilon^{-ax} \qquad (6b)$$

where ϵ is the base of the natural system of logarithms (2.71828) and a is the absorption coefficient. The spectral transmission curves for different thicknesses of the same glass are illustrated in Fig. 22.

It is often convenient to express the results of spectrophotometric measurements in terms of a quantity known as optical density. By definition, the optical density

$$D = \log_{10} \frac{1}{T}. \qquad (7)$$

From equation (6a), it will be seen that

$$D = xD_0 \qquad (8)$$

where D is the density of a material of thickness x, and D_0 is the density of a unit thickness of the same material. In other words, the density of a material is directly proportional to its thickness. The curves in Fig. 22 have been replotted on a density scale in Fig. 23.

It will be noticed that the curves in both Fig. 22 and Fig. 23 change their shape as the thickness of the sample varies. One of the useful applications of spectrophotometry is the identification of an unknown coloring material. This identification is often facilitated by plotting the spectrophotometric data in such a manner that the shape of the curve does not vary with the thickness of the sample. From equation (6b), it follows that

$$D = \log_{10} \frac{1}{T} = 0.4343ax. \qquad (9)$$

Whence,

$$\log_{10} D = \log_{10} (\log_{10} \frac{1}{T}) = \log_{10} 0.4343a + \log_{10} x. \qquad (10)$$

Now a (the absorption coefficient) varies with the wavelength, whereas x (the thickness of the sample) is independent of the wavelength. Hence, the first term on the right side of the above equation determines the shape of the curve, and variations in the thickness merely shift the curve vertically without altering its shape. The advantage of this plotting procedure is exemplified in Fig. 24, for the material whose spectral transmission curves for various thicknesses were plotted in Fig. 22.

22. Beer's Law

Many of the applications of spectrophotometry relate to the determination of the transmission factors of dye

[4] Bouguer set forth this law in 1729. It was rediscovered some years later by Lambert and, as a consequence, is frequently called Lambert's law of absorption.

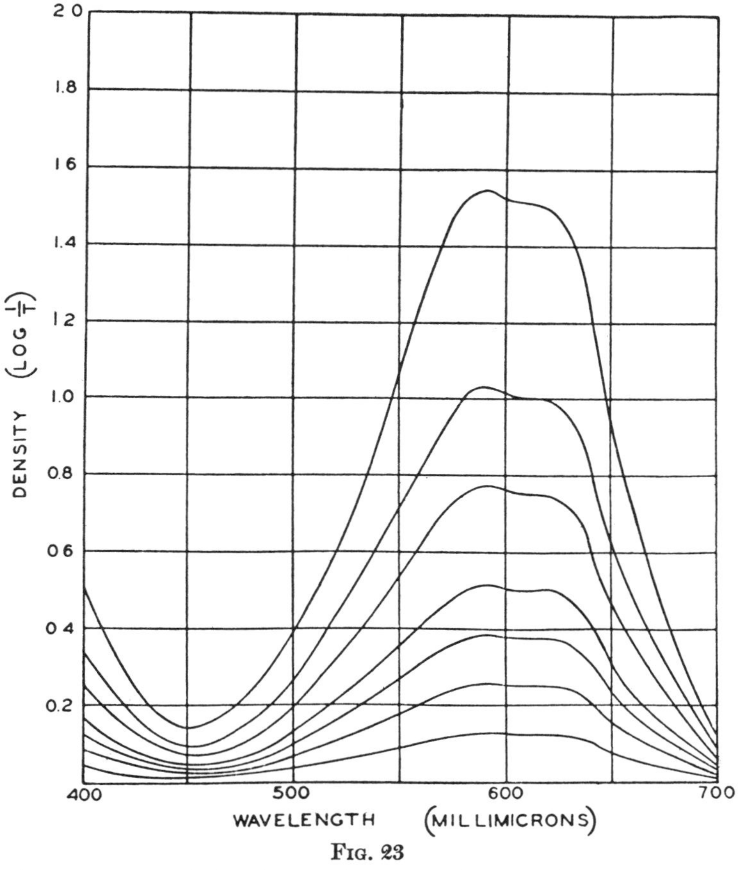

Fig. 23

Density curves of the glasses whose spectral transmission curves are shown in Fig. 22.

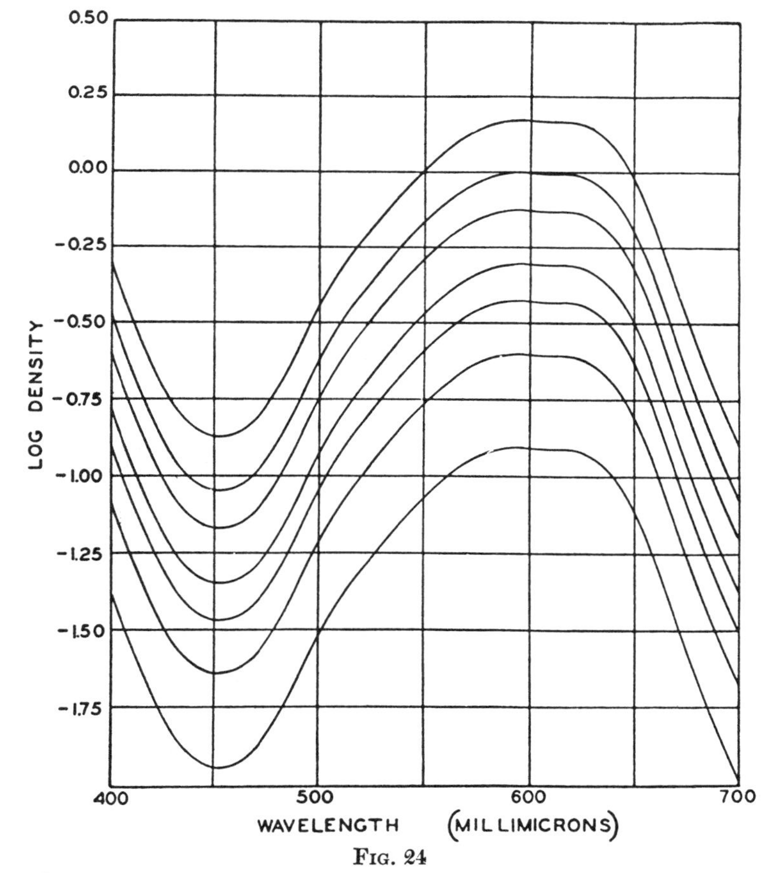

Fig. 24

Curves showing the logarithm of the densities of the glasses whose spectral transmission curves are shown in Fig. 22.

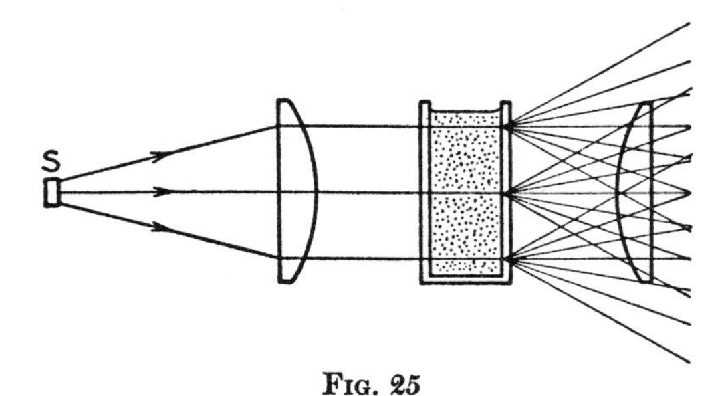

Fig. 25
Schematic diagram illustrating the unsuitability of the optical system shown in Fig. 21 for the measurement of non-homogeneous materials.

solutions. If c is the concentration of the dye, Beer's law states that the transmission factor of a sample of constant thickness is

$$T = t^c \qquad (11)$$

where t is the transmission factor of a solution of unit concentration. Combining Beer's law with Bouguer's law for samples of thickness x and concentration c

$$T = t^{cx} \qquad (12a)$$

where t is transmission factor of a sample of unit thickness and unit concentration. Corresponding to equation (6b), this expression may be written in the form[5]

$$T = \epsilon^{-acx}. \qquad (12b)$$

By analogy with equation (8), it is evident that the density of a definite thickness of dye solution is directly proportional to the concentration of the dye. The device of plotting the spectrophotometric curve of a dye solution in terms of log density again leads to a shape which is invariant with changes in either thickness or concentration. Bouguer's law is rigorously true in the case of homogeneous materials but Beer's law is often merely an approximation, especially at high concentrations.

23. Transmission Factors of Non-Homogeneous Materials

If an optically non-homogeneous material were inserted in the collimated beam shown in Fig. 21, the scattering of light that takes place within the material would give rise to the effect that is illustrated diagrammatically in Fig. 25. It is evident that much of this scattered light would not be admitted to the photometer. A value for the transmission factor at each wavelength could still be determined, but it would be significant only when the material is inserted in an optical system that is geometrically equivalent to that of the measuring instrument. It will be recalled in this connection that the transmission factor of homogeneous materials depends upon the length of the optical path and that, by convention, the light is made to travel parallel to the shortest dimension of the sample, thereby obtaining the maximum transmission. The maximum value of the transmission factors in the case of a non-homogeneous material is obtained if the optical system is so arranged as to collect all the light emerging from the material regardless of its direction. One method by which this is accomplished is illustrated in Fig. 26 where the sample is placed over the window of an integrating sphere. If the inside of the sphere is painted white, the same illumination on the window exposed to the photometer will result regardless of the direction in which the light enters the sphere. An arrangement like that of Fig. 26 is said to measure the diffuse transmission factor, whereas an arrangement like Fig. 21 is said to measure the specular transmission factor. Unless a material is almost perfectly homogeneous, it is usually more significant to determine its diffuse transmission factor. If a homogeneous material is illuminated by a collimated beam, the specular and diffuse transmission factors are equal. Hence, spectrophotometers are of more general usefulness if designed to measure the diffuse transmission factors of a sample.

[5] Equations (6b) and (12b) lead to different values of the constant unless a unit concentration is employed in connection with equation (6b).

24. Specular and Diffuse Reflection Factors

An opaque material whose surface is perfectly smooth behaves like a mirror. If, as shown in Fig. 27, a collimated beam of light is incident on a smooth surface at an angle α, the beam is still collimated after reflection, and makes an angle $-\alpha$ with the normal to the surface. The reflection factor of the surface at any wavelength is the ratio of the intensity of a monochromatic beam after reflection to its intensity before reflection. This type of reflection is called specular, from the Latin word speculum, meaning a mirror. The specular reflection factor varies somewhat with the angle of incidence, and a specification of this angle must accompany the result of the measurement. These measurements are not of much importance in the field of colorimetry, because materials exhibiting purely specular reflection are not often encountered.

In general, the surfaces of interest in colorimetry are rough in comparison with the wavelength of light. In this case, the reflected light is scattered in all directions,

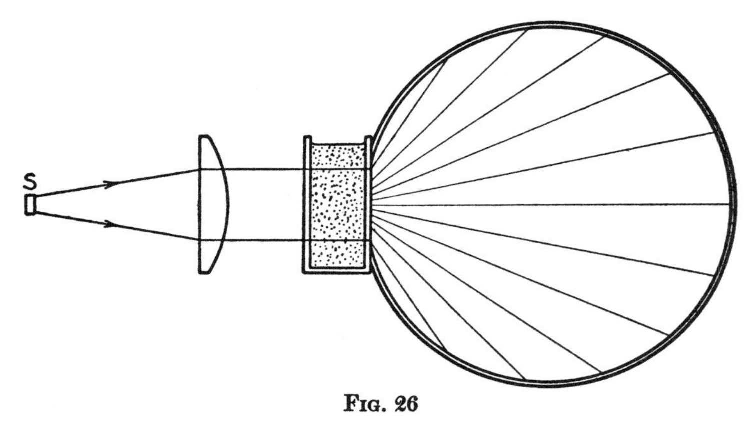

Fig. 26
Schematic diagram of an optical system designed for the measurement of the transmission factors of non-homogeneous materials.

and such surfaces are said to be diffusing. The procedure used in measuring the diffuse reflection factor of rough surfaces is illustrated diagrammatically in Fig. 28. A collimated beam illuminates the test sample, and the reflected light is collected by the walls of an integrating sphere. Due to an inherent property of the sphere, a window in the sphere wall is illuminated to the same extent by a given amount of light reflected from the test sample, regardless of the direction in which the light is reflected. Replacing the test sample by a white standard of known reflection factor enables the reflection factor of the former to be computed from the ratio of the brightness of the window in the two cases.[6]

The white standard that has obtained wide usage in colorimetry is a block of magnesium carbonate, one surface of which has been smoked with magnesium oxide. The smoking is accomplished by burning magnesium ribbon and allowing the white oxide to deposit on the surface until it has been coated with a layer of such thickness that further coatings do not change the reflection factor.[7]

It will be recalled that an instrument which measures the diffuse transmission of non-homogeneous materials can be used also to measure the specular transmission of homogeneous materials. In the same way, an arrangement like that shown in Fig. 28 may be used to obtain the specular reflection factors of a smooth surface, provided the precaution is taken to orient the sample in such a manner that the reflected beam strikes the sphere wall rather than the aperture through which the light enters the sphere. In this case it is especially necessary

[6] This method, although requiring a relatively simple optical system, is subject to an error that results from the fact that replacing the test sample by a white standard changes the average reflection factor of the walls of the sphere, even when the test sample occupies what would appear to be a negligible portion of the sphere wall. It is therefore better practice to use a direct comparison method rather than a substitution method. In the direct comparison methods, both the sample and the white standard occupy positions in the sphere wall simultaneously.

[7] Details of the preparation of such white standards are embodied in the Bureau of Standards *Letter Circular* No. 395, "Preparation and Colorimetric Properties of Magnesium-Oxide Reflectance Standard".

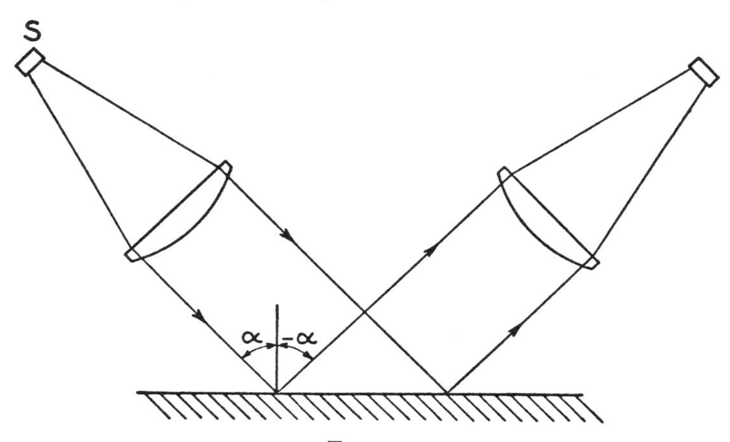

Fig. 27

Schematic diagram of a portion of the optical system employed in measuring the specular reflection factors of polished surfaces

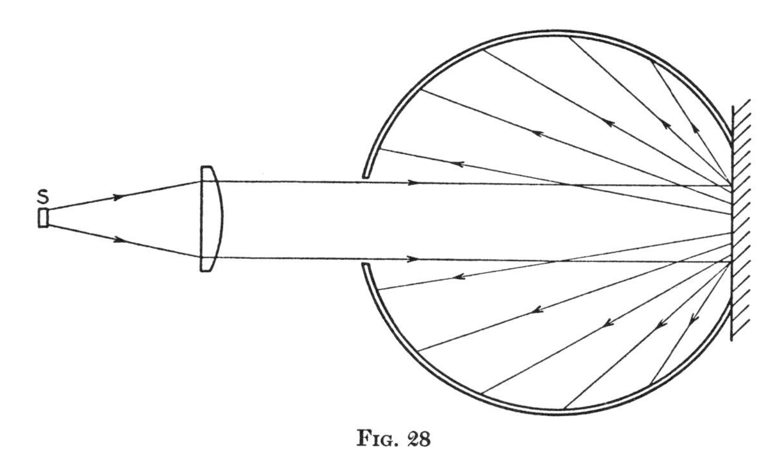

Fig. 28

Schematic diagram of an optical system designed for the measurement of the diffuse reflection factors of a rough surface

that the reflection factor of the white lining of the sphere be high.

It is evident that the methods outlined for the determination of the reflection factors of opaque materials are applicable to materials that are not opaque under the definition given in the first paragraph of this chapter. In the case of non-opaque materials, the reflection factor often depends to a considerable extent on the reflection factor of the material with which the sample is backed. Thus, if the test sample is backed with a perfectly white material, higher values are obtained than with a black backing. Since it is impossible to obtain a black material having a reflection factor of zero, the test sample may be supported, when desirable, in such a manner as to have no backing whatsoever. The light transmitted by the sample can then be absorbed by the walls of a darkened room. It may be added that many materials which are ordinarily assumed to be opaque have a measurable transmission factor. This is usually true of textile fabrics, paper samples, and films of paint, ink, lacquer, or enamel. For such samples, much valuable information concerning the opacity is acquired by measuring the diffuse reflection factors with both a white and a black backing. Tests of this sort will almost invariably demonstrate that the opacity is a function of wavelength, even in the case of nearly white materials.

25. Gloss

A reflecting surface that is perfectly diffusing is equally bright for all directions of viewing. This is true regardless of the mode of illumination. On the other hand, a perfect mirror illuminated by a single source of light appears bright for only a single direction of viewing, and black for all other directions. Neither type of surface can be fully realized in practice; all actual surfaces have diffusing characteristics that are intermediate between these two extremes. A surface that approaches closely the conditions of specular reflection is ordinarily said to be glossy; one that approaches more closely the conditions of diffuse reflection is said to be mat. Although the point was not emphasized in Chapter I, it is obvious that every color specification must be accompanied by a complete statement of the geometry of the illuminating beam and the geometry of that portion of the reflected (or transmitted) beam that is evaluated in the measurement.

There is no logical basis for the establishment of a preferred mode of illumination and observation. Indeed, colored materials are ordinarily examined under an almost infinite diversity of modes of illumination and viewing. Correspondingly, many modes of illumination and observation have been incorporated into the design of color measuring instruments. One design that has been used rather extensively employs diffuse illumination of the sample and observation of the sample by a nearly unidirectional beam normal to the surface. The color specification obtained with such an instrument corresponds to placing the sample horizontally where it is illuminated by the completely overcast sky, and viewing the sample from above. In the design of an instrument employing this mode of illumination and observation, it is frequently simpler to illuminate the sample normally by a unidirectional beam of light, and then to collect all the reflected light by means of an integrating sphere. A fundamental law of optics assures that this reversal of the direction of the light will yield identical results.

It may be mentioned here in passing that the reversibility of the light path has been assumed on numerous occasions throughout this volume. For example, in the description of the principles underlying the operation of a spectrophotometer, it was stated that the light incident on the sample is first dispersed into its spectral components. An equivalent procedure is to illuminate the sample by undispersed white light, and subsequently to disperse the reflected light into its spectral components.

CHAPTER IV

THE LAWS OF COLOR MIXTURE

Anyone who deals constantly with colors acquires in time the ability to predict the result of mixing the colors with which he is familiar. Knowledge acquired in this manner cannot be imparted to others, and is not applicable to new colors without performing an extended series of experiments with each. Fortunately, it is often possible to calculate the result of mixing colors from the characteristics of the components. This procedure, when valid, yields quantitative results of high precision, and is available alike to tyro and expert. The principles upon which such calculations are based are discussed in this chapter.

26. The Subtractive Method of Color Mixture

If two homogeneous materials, such as two pieces of glass or two cells containing clear liquids, are placed in series in the same beam of light, the transmission factor of the combination at any wavelength is the product of their individual transmission factors, neglecting inter-surface reflections. Hence, if the spectral transmission characteristics of two samples are known, the spectral transmission characteristics of the combination can be determined by computation. This has been done in Fig. 29 for a combination of a yellow and a blue glass. The resulting curve corresponds to a green glass as would be expected. It is evident that this principle may be generalized. Thus, if several materials are placed in series in the same beam of light, the transmission factor of the combination at any wavelength is the product of their individual transmission factors. It is frequently simpler in dealing with combinations like this to plot the results on a density scale. The density of the combination at any wavelength is the sum of the individual densities. Hence, the density curve of the combination can be readily obtained by graphical construction from the component curves.

In the case of mixtures of two or more colored solutions in the same cell, the spectral transmission curve of the mixture is obtained by taking the product at each wavelength of the individual transmission factors of the components.[1] It is necessary, of course, to correct the measured values of the transmission factors for reflection and transmission losses due to the cell, and it is also necessary to allow for differences in concentration that result from the mutual dilution on mixing.

The simple law of subtractive mixtures is seldom adequate for the computation of the spectral reflection characteristics of a mixture in the case of opaque or semi-opaque materials. A typical and very common problem of this sort arises in the case of mixtures of paints, inks, and other pigmented materials. These materials usually consist of a nearly transparent vehicle in which one or more insoluble pigments are present in a finely-divided state. In the case of certain printing inks, the pigment particles have nearly the same refractive index as the vehicle in which they are suspended. Such inks are therefore optically homogeneous, and the simple law of subtractive mixtures is applicable. In the application of this law, allowance must be made for the reflection at the air-vehicle boundary, for the reflection characteristics of the support to which the ink is applied, for the thickness of the film (the optical thickness is twice the actual thickness), and for the mutual dilution of the components on mixing.

A pigmented film in which the vehicle has the same refractive index as the pigment particles would be worthless as a paint because of its transparency. In other words, such a film would have no "hiding power." The hiding power, or opacity, of a pigmented film results from the innumerable reflections that take place at the pigment-vehicle interfaces, the magnitude of each individual reflec-

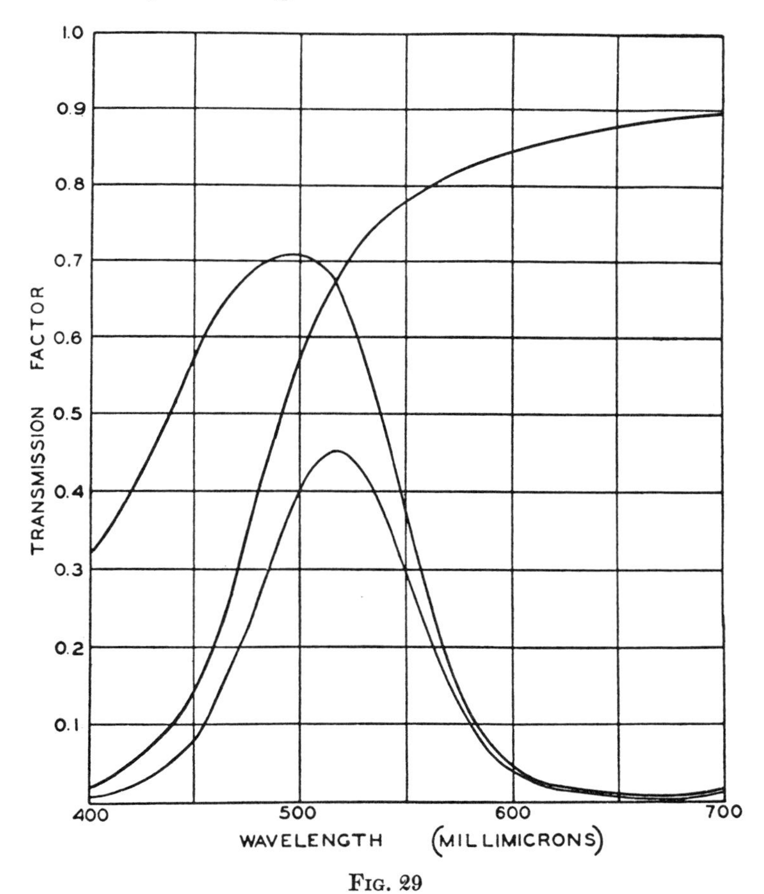

Fig. 29

Transmission curves of a blue glass, a yellow glass, and their combination.

[1] This law is rigorously true provided there is no interaction between the components.

tion depending on the difference between the two indices of refraction. Because the optical behavior of such materials is so complex, it is not to be expected that these materials will obey the simple law pertaining to homogeneous materials. Although attempts have been made to derive suitable formulae for the computation of the spectral reflection characteristics of such mixtures, it is usually safer and simpler to determine the mixture curves experimentally.

27. The Additive Method of Color Mixture

In the additive method of color mixture, the light from each component reaches the eye in an unmodified state. The simplest type of additive mixture is the simultaneous projection of two or more beams of colored light on the same area of a white screen. An equivalent procedure is to project the several beams, one at a time, in such rapid succession that they are fused by the eye into a single color. In the case of opaque materials such as colored papers, the same result is achieved by cutting the materials into such shapes that they can be mounted as segments of a circular disc. By spinning the disc about an axis through its center, the colors may be made to fuse. Such an arrangement is often called a color wheel or Maxwell disc. Another method of obtaining an additive mixture utilizes the limited resolving power of the eye. Thus, a roofing material composed of red and green particles may appear a dark yellow (brown) when viewed from such a distance that the individual particles can not be resolved. In the same way, a gray cloth is often made from a combination of black and white threads.

In the case where two or more beams of light are added together, as in the first example cited in the preceding paragraph, the energy distribution curve of the mixture is obtained by taking the sum of the energies of the components at each wavelength. In the other cases where one of the components replaces another, either in time or in space, allowance must be made for the proportion that each bears to the total time or total space. Thus, in the case of a Maxwell disc containing two colors occupying fractional areas a and b, the reflection factor of the rotating disc at each wavelength is obtained by multiplying the reflection factor of one component by a, the other component by b, and summing the result. This principle can be extended to include the addition of any number of colors.

It is unnecessary to have spectrophotometric data in order to compute the result of an additive mixture. Thus, if two beams of light whose tristimulus values are X_1, Y_1, Z_1 and X_2, Y_2, Z_2, respectively, are projected on the same white screen, the tristimulus values of the mixture are

$$X = X_1 + X_2 \qquad (13a)$$
$$Y = Y_1 + Y_2 \qquad (13b)$$
$$Z = Z_1 + Z_2 \qquad (13c)$$

If the colors of the two beams had been specified in terms of Y_1, x_1, y_1 and Y_2, x_2, y_2, the brightness and chromaticity of the mixture can be computed by means of the formulae

$$Y = Y_1 + Y_2 \qquad (13b)$$
$$x = \frac{m_1x_1 + m_2x_2}{m_1 + m_2} \qquad (14a)$$
$$y = \frac{m_1y_1 + m_2y_2}{m_1 + m_2} \qquad (14b)$$

where $m_1 = Y_1/y_1$ and $m_2 = Y_2/y_2$.

If two colors having fractional areas a and b are mixed on a Maxwell disc, the tristimulus values of the mixture are

$$X = aX_1 + bX_2 \qquad (15a)$$
$$Y = aY_1 + bY_2 \qquad (15b)$$
$$Z = aZ_1 + bZ_2 \qquad (15c)$$

If the two colors on the Maxwell disc had been expressed in terms of Y_1, x_1, y_1 and Y_2, x_2, y_2, the result of their mixture would be

$$Y = aY_1 + bY_2 \qquad (15b)$$
$$x = \frac{am_1x_1 + bm_2x_2}{am_1 + bm_2} \qquad (16a)$$
$$y = \frac{am_1y_1 + bm_2y_2}{am_1 + bm_2} \qquad (16b)$$

where again $m_1 = Y_1/y_1$ and $m_2 = Y_2/y_2$. The extension of these equations to mixtures of three or more colors is obvious.

Examination of Equations (14a) and (14b) shows that, regardless of the ratio in which two colors are mixed, the trichromatic coefficients, x, y, of the mixture are always such that the point thus determined on a chromaticity diagram lies on a straight line joining the points x_1, y_1 and x_2, y_2. This property of a chromaticity diagram was used extensively in Chapter I. In certain very common types of additive mixtures, the line joining the points representing the two components passes through the assumed white point. The dominant wavelengths of the components are then said to be complementary. Table VIII gives a list of pairs of complementary dominant wavelengths, assuming I.C.I. Illuminant C as the white point. If two materials have dominant wavelengths that are indicated in this table as complementary, they will, when mixed additively in the proper proportions, produce a mixture whose trichromatic coefficients are the same as those of Illuminant C.

Two materials whose dominant wavelengths are complementary under Illuminant C are not, in general, complementary under a different source of illumination. However, there are pairs of *physical complementaries* whose mixture is non-selective as to wavelength. Consider a color whose spectral reflection factor is R_1, where R_1 is understood to be a function of the wavelength. Let this be mixed additively in equal proportions with a color whose reflection factor is

$$R_2 = C - R_1$$

where C is any constant which results in R_2 being always greater than zero and less than one. Since these colors are mixed in equal proportions, the reflection factor of the mixture is

$$R = \tfrac{1}{2}R_1 + \tfrac{1}{2}C - \tfrac{1}{2}R_1 = \tfrac{1}{2}C.$$

Hence, this mixture matches a non-selective gray having a uniform reflection factor of $\frac{1}{2}C$ regardless of the type of illumination. It was shown in Chapter I that, by inverting the spectrophotometric curve of a color, the spectrophotometric curve of the complement is obtained. This is a special case corresponding to a value of $C = 1$. The resulting mixture then has a uniform reflection factor of 0.5. Other complementary pairs are theoretically possible within the limits set above for the value of C. The practical realization of these complementary pairs depends upon the availability of suitable pigments.

TABLE VIII

List of Complementary Wavelengths

Wavelength	*Complementary*	*Wavelength*	*Complementary*	*Wavelength*	*Complementary*
380 mμ	567.0 mμ	489 mμ	600.9 mμ	590 mμ	485.9 mμ
400	567.1			591	486.3
420	567.3	490	607.0	592	486.7
430	567.5	491	616.8	593	487.0
440	568.0	492	640.2	594	487.3
450	568.9			595	487.6
455	569.6	568	439.3	596	487.9
460	570.4	569	450.7	597	488.1
		570	457.9	598	488.4
470	573.1	571	463.1	599	488.6
471	573.6	572	466.8		
472	574.0	573	469.7	600	488.8
473	574.5	574	471.9	605	489.7
474	575.1	575	473.8	610	490.4
475	575.7	576	475.4	615	490.9
476	576.4	577	476.7	620	491.2
477	577.2	578	478.0	625	491.5
478	578.0	579	479.0	630	491.7
479	579.0				
		580	480.0	640	492.0
480	580.0	581	480.8	650	492.2
481	581.2	582	481.6	660	492.3
482	582.6	583	482.3	670	492.3
483	584.1	584	482.9	680	492.4
484	585.8	585	483.5	690	492.4
485	587.8	586	484.1	700	492.4
486	590.2	587	484.6	780	492.4
487	593.0	588	485.1		
488	596.5	589	485.5		

CHAPTER V

DETERMINATION OF TRISTIMULUS VALUES BY THE WEIGHTED ORDINATE METHOD

As explained in Chapter I, the objective method of evaluating the appearance of a color to a human observer consists in determining a stimulus that is equivalent under the assumed conditions. The stimulus ordinarily employed is a mixture of three basic or primary stimuli. In the calculation of the equivalent stimulus from spectrophotometric data, it is necessary to make use of the tristimulus values for the spectrum colors. The experimental procedure that is followed in determining these tristimulus values was outlined on page 8. Table I listed the values of $\bar{x}$, $\bar{y}$, $\bar{z}$ that were adopted by the International Commission on Illumination at the 1931 meeting. These values indicate the amount of each of the chosen primaries that would be required by a normal observer to color match equal amounts of energy at the several wavelengths.

28. Tristimulus Values of the Spectrum Colors

The experiments that led to the adoption of the basic data concerning the chromatic properties of the normal human eye were carried out at wavelength intervals of 10 millimicrons throughout the visible spectrum. The resolutions adopted at the 1931 meeting included these values (after adjustment to a more convenient set of primaries) and also included the values obtained by interpolation at intervals of 5 millimicrons. For many purposes, and especially for the construction of auxiliary tables, it is desirable to have a somewhat closer interpolation. Table XII contains the results of such an interpolation using the third difference osculatory formula.

The detailed steps employed in carrying out this interpolation were as follows: The values of the trichromatic coefficients x and z were taken directly from the I.C.I. report, and were interpolated to wavelength intervals of 1 millimicron by the third difference osculatory formulae. The value of y at each wavelength was then computed by means of the equation

$$y = 1 - (x + z).$$

The $\bar{y}$ function had previously been interpolated by Judd[1] to intervals of 1 millimicron. The values of $\bar{x}$ and $\bar{z}$ were determined by means of the following equations:

$$\bar{x} = (\bar{y}/y)x$$

$$\bar{z} = (\bar{y}/y)z.$$

[1] *Bureau of Standards Journal of Research*, v. 6, p. 465 (1931). The values listed in Table XII were supplied by Dr. Deane B. Judd from his original calculations.

The values of x, y, and z recommended by the International Commission on Illumination were given to four significant figures. These values do not change significantly at wavelengths longer than 700 millimicrons. Judd had previously carried the interpolation of $\bar{y}$ to a greater number of decimal places; correspondingly, the values of $\bar{x}$ and $\bar{z}$ are listed in Table XII to seven decimal places. As in some of the previous tables, these additional figures are separated from the significant figures by a space. Table XII is completely consistent with the I.C.I. values of x, z, and $\bar{y}$. The few slight discrepancies that may be noticed between Table I and Table XII are the result of accepting the I.C.I. recommendations for x, z and $\bar{y}$ as the bases for interpolation. Corresponding inconsistencies, not entirely attributable to rejection errors, are present between the values of x, y, z and $\bar{x}$, $\bar{y}$, $\bar{z}$ in the original I.C.I. recommendations.

As an example of the use of these tristimulus values, consider the problem of determining the trichromatic coefficients for Illuminant C. The distribution of energy in this source is known from Table V of Chapter II. These values are set down in column two of Table IX against the corresponding wavelength entries in column one. To save space, the wavelength interval selected is 10 millimicrons. In general, the wavelength intervals should be so small that the use of smaller intervals would not produce significant changes in the final results. Column three in the table gives the corresponding values of $\bar{x}$ from Table XII; column four, the product of columns two and three. The sum of the entries in the fourth column is a measure of X, one of the tristimulus values for Illuminant C. A repetition of this procedure with $\bar{y}$ and $\bar{z}$ substituted in turn for $\bar{x}$ gives the following results:

$$X = 1044$$
$$Y = 1064$$
$$Z = 1257$$

It will be noticed that the values in column two give only the relative amount of energy at each wavelength, not the absolute amount. Furthermore, the computed values of X, Y, and Z are the sums of entries corresponding to intervals of 10 millimicrons, and different sums would have been obtained if a different wavelength interval had been selected. Fortunately, the vast majority of problems in colorimetry are concerned with the quality of the light source, not with the quantity of light that it emits. Multiplying X, Y, and Z by a common factor does not alter the specification of quality which is evaluated by

$$x = X/(X + Y + Z)$$
$$y = Y/(X + Y + Z).$$

Substitution in these equations in the above case gives $x = 0.3101$, $y = 0.3163$. These are the trichromatic coefficients of Illuminant C.

TABLE IX

SAMPLE CALCULATION OF A TRISTIMULUS VALUE OF A SOURCE BY THE WEIGHTED ORDINATE METHOD

Wavelength	E_C	$\bar{x}$	$E_C\bar{x}$
380 mμ	33.0	0.0014	0.05
390	47.4	0.0043	0.20
400	63.3	0.0144	0.91
410	80.6	0.0432	3.48
420	98.1	0.1344	13.19
430	112.4	0.2839	31.92
440	121.5	0.3469	42.15
450	124.0	0.3362	41.69
460	123.1	0.2909	35.81
470	123.8	0.1954	24.19
480	123.9	0.0956	11.85
490	120.7	0.0320	3.86
500	112.1	0.0049	0.55
510	102.3	0.0093	0.95
520	96.9	0.0633	6.13
530	98.0	0.1655	16.22
540	102.1	0.2904	29.65
550	105.2	0.4335	45.60
560	105.3	0.5945	62.60
570	102.3	0.7622	77.97
580	97.8	0.9162	89.61
590	93.2	1.0266	95.68
600	89.7	1.0620	95.26
610	88.4	1.0028	88.65
620	88.1	0.8545	75.28
630	88.0	0.6424	56.54
640	87.8	0.4479	39.32
650	88.2	0.2835	25.01
660	87.9	0.1649	14.50
670	86.3	0.0874	7.54
680	84.0	0.0468	3.93
690	80.2	0.0227	1.82
700	76.3	0.0114	0.87
710	72.4	0.0058	0.42
720	68.3	0.0029	0.20
730	64.4	0.0014	0.09
740	61.5	0.0007	0.04
750	59.2	0.0003	0.02
760	58.1	0.0002	0.01
770	58.2	0.0001	0.01
		Sum =	1043.8

29. TRISTIMULUS VALUES OF THE SPECTRUM COLORS WEIGHTED BY THE I.C.I. ILLUMINANTS

Most of the problems of importance in colorimetry relate to the determination of tristimulus values of either transparent or opaque materials. Such problems far outnumber those that relate exclusively to light sources. The tristimulus values for either a transparent or an opaque material may be computed by multiplying at each wavelength the relative amount of energy in the source by the reflection factor or transmission factor of the sample. The product gives the relative amount of energy at this wavelength entering the eye of the observer. This product must then be multiplied by $\bar{x}$, $\bar{y}$, and $\bar{z}$ in turn. It is clear that an equivalent procedure is to multiply the amount of energy in any source that is to be used extensively by the corresponding values of $\bar{x}$, $\bar{y}$, $\bar{z}$. With values of such products at each wavelength, it is then only necessary to multiply by the reflection factors or the transmission factors of the sample. Tables XIII, XIV, and XV list values of $E\bar{x}$, $E\bar{y}$, and $E\bar{z}$ for the three I.C.I. illuminants. These values were obtained by multiplying at each wavelength values taken from the appropriate tables of Chapter II by the corresponding values of $\bar{x}$, $\bar{y}$, and $\bar{z}$ given in Table XII of this chapter. As before, the figures following the space are to be used only for the compilation of other tables and should ordinarily be ignored.

As an example of the use of these tables, consider the problem of determining the trichromatic coefficients and visual transmission of a red signal glass which is to be placed in front of a lamp having the quality of Illuminant A. The computational procedure by which these results are obtained is outlined in Table X. Column one gives the wavelength of the light at intervals of 10 millimicrons; column two gives the transmission factors of the red glass as determined by a spectrophotometer and shown in Fig. 20; column three gives the values of $E\bar{x}$ for Illuminant A from Table XIII; and column four is the product of columns two and three. The sum of the entries in the fourth column is a measure of X. Repeating this operation with $E\bar{y}$ and $E\bar{z}$ substituted for $E\bar{x}$ gives $X = 640.2$, $Y = 321.3$, and $Z = 0.0$. From these values, the trichromatic coefficients are $x = 0.6658$ and $y = 0.3342$.

The visual transmission can now be found by determining the ratio of the value of Y to the value of Y for a hypothetical glass whose transmission factor is 1.000 at every wavelength. The value of Y obtained for such a hypothetical glass is found to be 1079.0, whereas the value obtained for the red glass is 321.3. The ratio

$$\frac{321.3}{1079.0} = 0.2978 = 29.78\%.$$

It will be recalled that this procedure is valid because the $\bar{y}$ function was made by appropriate selection of the primaries to correspond to the so-called visibility function of the eye. The above ratio is, therefore, the ratio of the number of lumens emitted by the lamp and signal glass combination to the number of lumens that would be emitted by the lamp alone. This ratio is sometimes called the visual efficiency of the glass. It should be remembered that the visual efficiency of a filter depends on the spectral quality of the source with which it is used.

As a second example in the use of Tables XIII, XIV, and XV, consider the problem of evaluating the color of a nearly white paper whose reflection factors at the various wavelengths are listed in column two of Table XI, and shown in Fig. 30. In this table, column 1 gives the wavelength at intervals of 10 millimicrons, an interval which is sufficiently small in dealing with nearly white substances. The third column in the table gives the values of $E\bar{x}$ for Illuminant C from Table XV. The fourth column gives the products of the entries in columns two and three. The sum of the entries in the fourth column is a measure of X. Repeating this operation with $E\bar{y}$ and $E\bar{z}$ substituted in turn for $E\bar{x}$ gives $X = 880$, $Y = 902$, and $Z = 960$. From these, the trichromatic coefficients are $x = 0.3209, y = 0.3290$.

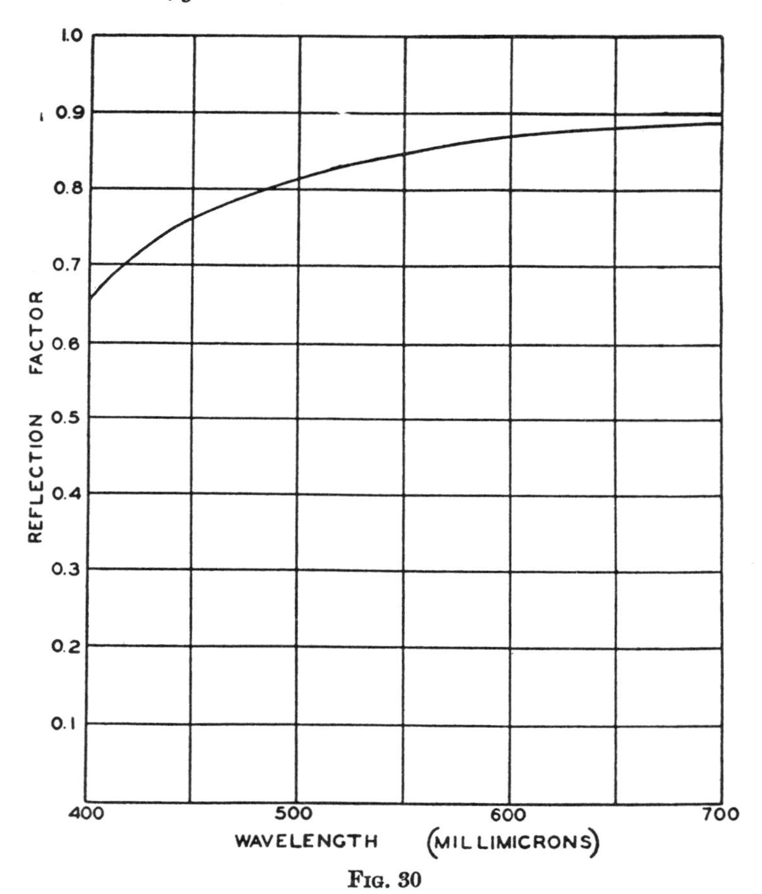

FIG. 30
Diffuse reflection curve of a sample of "white" paper.

The term visual efficiency is equally applicable to reflecting surfaces, although it has not been used to any great extent. The term brightness is in common use, although this term has been used in so many senses that it is unquestionably better practice to use the term visual efficiency. As in the previous example, the visual efficiency is computed by determining the value of Y for a surface having a reflection factor of 1.000 at all wavelengths. The resulting value of Y for such a surface is 1065. Hence the ratio

$$\frac{902}{1065} = 0.847 = 84.7\%$$

which is the visual efficiency of the white paper. In other words, it is the brightness of the white paper relative to the brightness of a perfect reflector under the same illumination, in this case under Illuminant C. As before, this visual efficiency, or relative brightness, depends on the spectral quality of the illumination.

TABLE X

SAMPLE CALCULATION OF A TRISTIMULUS VALUE OF A FILTER BY THE WEIGHTED ORDINATE METHOD

Wavelength	T	$E_A\bar{x}$	$TE_A\bar{x}$
380 mμ	.....	.....	....
...	.....	.....	....
...	.....	.....	....
...	.....	.....	....
570	0.002	81.69	0.2
580	0.100	104.85	10.5
590	0.500	124.97	62.5
600	0.775	137.04	106.2
610	0.850	136.73	116.2
620	0.870	122.72	106.8
630	0.876	96.90	84.9
640	0.880	68.20	60.0
650	0.883	44.66	39.4
660	0.887	26.81	23.8
670	0.890	15.63	13.9
680	0.894	8.67	7.8
690	0.895	4.35	3.9
700	0.897	2.25	2.0
710	0.897	1.19	1.1
720	0.897	0.61	0.5
730	0.897	0.31	0.3
740	0.897	0.15	0.1
750	0.897	0.08	0.1
760	0.897	0.04	0.0
770	0.897	0.02	0.0
			Sum = 640.2

TABLE XI

Sample Calculation of a Tristimulus Value for a Sheet of "White" Paper by the Weighted Ordinate Method

Wavelength	R	$E_C\bar{x}$	$RE_C\bar{x}$	*Wavelength*	R	$E_C\bar{x}$	$RE_C\bar{x}$
380 mμ	0.620	0.05	0.0	570 mμ	0.855	77.97	66.7
390	0.640	0.20	0.1	580	0.860	89.61	77.1
				590	0.865	95.68	82.8
400	0.660	0.91	0.6				
410	0.681	3.48	2.4	600	0.870	95.26	82.9
420	0.703	13.19	9.3	610	0.873	88.65	77.4
430	0.727	31.92	23.2	620	0.875	75.28	65.9
440	0.745	42.15	31.4	630	0.878	56.54	49.6
450	0.760	41.69	31.7	640	0.880	39.32	34.6
460	0.775	35.81	27.8	650	0.882	25.01	22.1
470	0.787	24.19	19.0	660	0.884	14.50	12.8
480	0.800	11.85	9.5	670	0.884	7.54	6.7
490	0.805	3.86	3.1	680	0.885	3.93	3.5
				690	0.885	1.82	1.6
500	0.815	0.55	0.4				
510	0.822	0.95	0.8	700	0.885	0.87	0.8
520	0.830	6.13	5.1	710	0.885	0.42	0.4
530	0.835	16.22	13.5	720	0.885	0.20	0.2
540	0.840	29.65	24.9	730	0.885	0.09	0.1
550	0.845	45.60	38.5	740	0.885	0.04	0.0
560	0.850	62.60	53.2				
							Sum = 880.0

TABLE XII

Tristimulus Values for Spectrum Colors

(Equal Energy)

Wavelength	$\bar{x}$	$\bar{y}$	$\bar{z}$	*Wavelength*	$\bar{x}$	$\bar{y}$	$\bar{z}$
380 mμ	0.0013 928	0.0000 4000	0.0065 672	410 mμ	0.0431 500	0.0012 000	0.2056 500
81	0.0015 697	0.0000 4499	0.0074 024	11	0.0495 395	0.0013 774	0.2362 519
82	0.0017 293	0.0000 4947	0.0081 563	12	0.0560 145	0.0015 552	0.2673 074
83	0.0018 869	0.0000 5388	0.0089 013	13	0.0628 005	0.0017 418	0.2998 997
84	0.0020 580	0.0000 5866	0.0097 103	14	0.0700 799	0.0019 456	0.3349 084
385	0.0022 582	0.0000 6425	0.0106 573	415	0.0779 828	0.0021 750	0.3729 671
86	0.0025 035	0.0000 7110	0.0118 176	16	0.0867 437	0.0024 384	0.4152 030
87	0.0028 098	0.0000 7966	0.0132 671	17	0.0966 553	0.0027 442	0.4630 287
88	0.0031 932	0.0000 9037	0.0150 815	18	0.1078 562	0.0031 008	0.5171 422
89	0.0036 699	0.0001 0367	0.0173 376	19	0.1204 365	0.0035 166	0.5780 190
390	0.0042 563	0.0001 2000	0.0201 134	420	0.1344 313	0.0040 000	0.6458 823
91	0.0048 911	0.0001 376	0.0231 200	21	0.1495 296	0.0045 485	0.7192 897
92	0.0055 188	0.0001 549	0.0260 952	22	0.1652 304	0.0051 520	0.7958 591
93	0.0061 747	0.0001 729	0.0292 059	23	0.1812 517	0.0058 075	0.8742 528
94	0.0068 936	0.0001 926	0.0326 172	24	0.1973 335	0.0065 120	0.9532 262
395	0.0077 106	0.0002 150	0.0364 958	425	0.2132 420	0.0072 625	1.0316 506
96	0.0086 577	0.0002 410	0.0409 936	26	0.2288 987	0.0080 560	1.1091 219
97	0.0097 681	0.0002 715	0.0460 694	27	0.2441 368	0.0088 895	1.1848 456
98	0.0110 780	0.0003 075	0.0524 954	28	0.2586 113	0.0097 600	1.2572 007
99	0.0126 239	0.0003 500	0.0598 456	29	0.2719 858	0.0106 645	1.3246 108
400	0.0144 416	0.0004 000	0.0684 916	430	0.2839 478	0.0116 000	1.3856 115
1	0.0162 578	0.0004 499	0.0771 372	31	0.2945 788	0.0125 699	1.4406 126
2	0.0178 909	0.0004 947	0.0849 213	32	0.3041 523	0.0135 792	1.4908 992
3	0.0194 972	0.0005 388	0.0925 856	33	0.3126 842	0.0146 273	1.5365 136
4	0.0212 326	0.0005 866	0.1008 717	34	0.3202 068	0.0157 136	1.5775 719
405	0.0232 523	0.0006 425	0.1105 194	435	0.3267 649	0.0168 375	1.6142 463
6	0.0257 113	0.0007 110	0.1222 677	36	0.3324 167	0.0179 984	1.6467 441
7	0.0287 685	0.0007 966	0.1368 759	37	0.3372 136	0.0191 957	1.6753 116
8	0.0325 846	0.0009 037	0.1551 563	38	0.3411 980	0.0204 288	1.7002 221
9	0.0373 227	0.0010 367	0.1777 709	39	0.3444 126	0.0216 971	1.7217 559

TABLE XII (*Continued*)

Wavelength	$\bar{x}$	$\bar{y}$	$\bar{z}$	*Wavelength*	$\bar{x}$	$\bar{y}$	$\bar{z}$
440 mμ	0.3468 990	0.0230 000	1.7401 926	500 mμ	0.0049 194	0.3230 000	0.2720 063
41	0.3487 178	0.0243 227	1.7557 491	1	0.0038 519	0.3382 4	0.2591 047
42	0.3498 426	0.0256 576	1.7682 066	2	0.0030 216	0.3544 1	0.2467 062
43	0.3501 923	0.0270 149	1.7772 284	3	0.0024 724	0.3713 8	0.2347 692
44	0.3497 007	0.0284 048	1.7825 468	4	0.0022 546	0.3890 2	0.2232 784
445	0.3483 203	0.0298 375	1.7839 798	505	0.0024 255	0.4072 5	0.2122 700
46	0.3463 094	0.0313 232	1.7828 263	6	0.0029 907	0.4259 4	0.2013 803
47	0.3440 126	0.0328 721	1.7807 567	7	0.0039 315	0.4449 6	0.1903 409
48	0.3415 162	0.0344 944	1.7781 659	8	0.0052 774	0.4642 3	0.1793 412
49	0.3388 931	0.0362 003	1.7753 854	9	0.0070 612	0.4836 2	0.1685 551
450	0.3362 033	0.0380 000	1.7726 892	510	0.0093 197	0.5030 0	0.1581 680
51	0.3335 710	0.0398 76	1.7707 005	11	0.0121 423	0.5228 6	0.1481 619
52	0.3308 409	0.0418 08	1.7686 277	12	0.0155 692	0.5435 5	0.1384 022
53	0.3277 023	0.0438 02	1.7648 451	13	0.0195 690	0.5648 4	0.1289 774
54	0.3239 129	0.0458 64	1.7580 485	14	0.0241 065	0.5865 0	0.1200 182
455	0.3192 951	0.0480 00	1.7472 423	515	0.0291 390	0.6082 5	0.1116 872
56	0.3142 879	0.0502 16	1.7347 569	16	0.0347 775	0.6298 6	0.1040 558
57	0.3091 865	0.0525 18	1.7221 450	17	0.0411 136	0.6511 0	0.0969 878
58	0.3037 113	0.0549 12	1.7077 884	18	0.0480 541	0.6716 9	0.0903 829
59	0.2976 616	0.0574 04	1.6904 601	19	0.0554 854	0.6914 0	0.0841 625
460	0.2909 090	0.0600 00	1.6692 929	520	0.0632 681	0.7100 0	0.0782 549
61	0.2841 365	0.0626 59	1.6479 516	21	0.0715 018	0.7277 0	0.0729 586
62	0.2774 155	0.0653 52	1.6268 527	22	0.0803 493	0.7448 8	0.0684 625
63	0.2699 541	0.0681 03	1.6015 361	23	0.0897 299	0.7615 4	0.0645 207
64	0.2612 093	0.0709 36	1.5688 086	24	0.0995 352	0.7776 4	0.0608 669
465	0.2508 787	0.0738 75	1.5267 500	525	0.1096 372	0.7931 9	0.0572 187
66	0.2398 662	0.0769 44	1.4805 436	26	0.1200 907	0.8081 6	0.0536 976
67	0.2289 557	0.0801 67	1.4346 692	27	0.1310 013	0.8225 4	0.0505 525
68	0.2179 478	0.0835 68	1.3877 093	28	0.1422 778	0.8363 2	0.0476 654
69	0.2067 569	0.0871 71	1.3389 036	29	0.1538 101	0.8494 8	0.0449 081
470	0.1953 823	0.0910 00	1.2880 121	530	0.1654 689	0.8620 0	0.0421 426
71	0.1842 995	0.0950 17	1.2380 615	31	0.1772 173	0.8738 6	0.0394 156
72	0.1736 659	0.0991 76	1.1901 553	32	0.1891 168	0.8850 6	0.0368 533
73	0.1631 896	0.1034 89	1.1422 855	33	0.2011 746	0.8956 2	0.0344 271
74	0.1527 199	0.1079 68	1.0932 982	34	0.2134 012	0.9055 7	0.0321 048
475	0.1422 085	0.1126 25	1.0426 895	535	0.2258 085	0.9149 4	0.0298 502
76	0.1320 182	0.1174 72	0.9930 107	36	0.2383 741	0.9237 5	0.0277 000
77	0.1223 752	0.1225 21	0.9459 877	37	0.2510 887	0.9320 4	0.0257 026
78	0.1131 427	0.1277 84	0.9006 378	38	0.2639 741	0.9398 2	0.0238 326
79	0.1042 423	0.1332 73	0.8563 639	39	0.2770 654	0.9471 4	0.0220 576
480	0.0956 345	0.1390 00	0.8128 409	540	0.2903 863	0.9540 0	0.0203 624
81	0.0872 940	0.1448 425	0.7704 779	41	0.3039 207	0.9603 5	0.0187 537
82	0.0792 756	0.1507 200	0.7297 207	42	0.3176 231	0.9661 2	0.0172 701
83	0.0716 681	0.1567 075	0.6904 842	43	0.3314 922	0.9713 4	0.0158 954
84	0.0645 432	0.1628 800	0.6527 085	44	0.3455 272	0.9760 4	0.0146 113
485	0.0579 559	0.1693 125	0.6163 413	545	0.3597 272	0.9802 5	0.0133 984
86	0.0518 619	0.1760 800	0.5816 122	46	0.3740 954	0.9840 0	0.0122 818
87	0.0462 132	0.1832 575	0.5490 127	47	0.3886 425	0.9873 2	0.0112 855
88	0.0410 262	0.1909 200	0.5187 237	48	0.4033 795	0.9902 4	0.0103 851
89	0.0362 971	0.1991 425	0.4907 704	49	0.4183 182	0.9927 9	0.0095 548
490	0.0320 108	0.2080 000	0.4650 738	550	0.4334 710	0.9950 0	0.0087 671
91	0.0279 982	0.2173 445	0.4404 717	51	0.4488 287	0.9968 54	0.0080 380
92	0.0241 846	0.2270 160	0.4164 714	52	0.4643 704	0.9983 12	0.0073 913
93	0.0206 620	0.2370 715	0.3937 035	53	0.4800 909	0.9993 68	0.0068 097
94	0.0174 930	0.2475 680	0.3725 512	54	0.4959 839	1.0000 16	0.0062 752
495	0.0147 230	0.2585 625	0.3532 288	555	0.5120 417	1.0002 50	0.0057 686
96	0.0122 508	0.2701 120	0.3350 687	56	0.5282 710	1.0000 64	0.0052 999
97	0.0099 875	0.2822 735	0.3175 979	57	0.5446 633	0.9994 52	0.0048 854
98	0.0079 824	0.2951 040	0.3011 326	58	0.5611 855	0.9984 08	0.0045 141
99	0.0062 788	0.3086 605	0.2858 863	59	0.5778 010	0.9969 26	0.0041 748

TABLE XII (*Continued*)

Wavelength	$\bar{x}$	$\bar{y}$	$\bar{z}$	*Wavelength*	$\bar{x}$	$\bar{y}$	$\bar{z}$
560 mμ	0.5944 697	0.9950 0	0.0038 558	620 mμ	0.8544 512	0.3810 0	0.0001 977
61	0.6111 757	0.9926 2	0.0035 631	21	0.8354 079	0.3690 4	0.0001 847
62	0.6279 195	0.9897 8	0.0033 068	22	0.8153 923	0.3570 0	0.0001 727
63	0.6446 892	0.9864 9	0.0030 802	23	0.7945 991	0.3449 4	0.0001 612
64	0.6614 790	0.9827 7	0.0028 765	24	0.7732 255	0.3329 2	0.0001 502
565	0.6782 764	0.9786 3	0.0026 885	625	0.7514 705	0.3210 0	0.0001 394
66	0.6950 933	0.9740 7	0.0025 217	26	0.7294 592	0.3092 4	0.0001 288
67	0.7119 493	0.9691 3	0.0023 821	27	0.7073 310	0.2977 0	0.0001 186
68	0.7287 694	0.9637 8	0.0022 631	28	0.6853 065	0.2864 4	0.0001 088
69	0.7455 418	0.9580 8	0.0021 580	29	0.6636 049	0.2755 2	0.0000 995
570	0.7621 835	0.9520 0	0.0020 595	630	0.6424 434	0.2650 0	0.0000 907
71	0.7786 744	0.9455 2	0.0019 705	31	0.6217 695	0.2548 5	0.0000 822
72	0.7949 794	0.9385 9	0.0018 965	32	0.6013 586	0.2449 8	0.0000 740
73	0.8110 812	0.9312 5	0.0018 334	33	0.5812 653	0.2354 0	0.0000 664
74	0.8269 269	0.9235 0	0.0017 774	34	0.5614 199	0.2260 7	0.0000 593
575	0.8424 972	0.9153 8	0.0017 244	635	0.5418 738	0.2170 0	0.0000 531
76	0.8577 987	0.9069 0	0.0016 770	36	0.5225 924	0.2081 7	0.0000 478
77	0.8728 726	0.8981 1	0.0016 384	37	0.5035 396	0.1995 6	0.0000 431
78	0.8876 586	0.8890 1	0.0016 048	38	0.4847 643	0.1911 8	0.0000 389
79	0.9021 332	0.8796 4	0.0015 725	39	0.4661 875	0.1829 9	0.0000 349
580	0.9162 317	0.8700 0	0.0015 374	640	0.4478 554	0.1750 0	0.0000 311
81	0.9298 795	0.8600 2	0.0014 997	41	0.4297 833	0.1672 1	0.0000 275
82	0.9429 475	0.8495 9	0.0014 617	42	0.4120 346	0.1596 4	0.0000 240
83	0.9554 255	0.8387 8	0.0014 235	43	0.3946 254	0.1522 9	0.0000 208
84	0.9673 069	0.8276 6	0.0013 853	44	0.3775 709	0.1451 6	0.0000 178
585	0.9785 300	0.8162 5	0.0013 470	645	0.3608 865	0.1382 5	0.0000 150
86	0.9891 925	0.8046 2	0.0013 092	46	0.3445 786	0.1315 6	0.0000 123
87	0.9993 825	0.7928 4	0.0012 720	47	0.3286 626	0.1250 9	0.0000 098
88	1.0090 209	0.7809 3	0.0012 350	48	0.3131 655	0.1188 4	0.0000 075
89	1.0180 922	0.7689 6	0.0011 978	49	0.2981 112	0.1128 1	0.0000 055
590	1.0265 858	0.7570 0	0.0011 600	650	0.2835 212	0.1070 0	0.0000 039
91	1.0343 893	0.7449 4	0.0011 216	51	0.2694 822	0.1014 4	0.0000 026
92	1.0413 578	0.7326 6	0.0010 828	52	0.2559 709	0.0961 2	0.0000 015
93	1.0474 444	0.7202 2	0.0010 435	53	0.2430 273	0.0910 5	0.0000 007
94	1.0525 317	0.7076 3	0.0010 038	54	0.2305 836	0.0862 0	0.0000 002
595	1.0565 611	0.6949 4	0.0009 638	655	0.2186 240	0.0815 6	
96	1.0594 778	0.6821 7	0.0009 235	56	0.2071 283	0.0771 2	
97	1.0613 756	0.6693 6	0.0008 831	57	0.1960 476	0.0728 6	
98	1.0623 510	0.6565 4	0.0008 426	58	0.1853 344	0.0687 6	
99	1.0625 072	0.6437 4	0.0008 023	59	0.1749 947	0.0648 2	
600	1.0619 702	0.6310 0	0.0007 621	660	0.1649 259	0.0610 0	
1	1.0606 587	0.6182 5	0.0007 222	61	0.1552 634	0.0573 5	
2	1.0583 988	0.6054 3	0.0006 824	62	0.1460 675	0.0538 9	
3	1.0551 249	0.5925 6	0.0006 428	63	0.1372 967	0.0506 0	
4	1.0507 750	0.5796 6	0.0006 034	64	0.1290 158	0.0475 0	
605	1.0452 905	0.5667 5	0.0005 644	665	0.1211 521	0.0445 6	
6	1.0386 889	0.5538 6	0.0005 244	66	0.1137 070	0.0417 8	
7	1.0310 697	0.5410 2	0.0004 837	67	0.1066 241	0.0391 4	
8	1.0225 039	0.5282 5	0.0004 442	68	0.0998 779	0.0366 3	
9	1.0130 470	0.5155 6	0.0004 079	69	0.0934 976	0.0342 6	
610	1.0028 369	0.5030 0	0.0003 765	670	0.0874 030	0.0320 0	
11	0.9917 737	0.4905 3	0.0003 508	71	0.0817 308	0.0299 0	
12	0.9797 790	0.4781 2	0.0003 294	72	0.0765 925	0.0280 0	
13	0.9668 992	0.4657 7	0.0003 107	73	0.0719 371	0.0262 8	
14	0.9531 817	0.4534 8	0.0002 934	74	0.0676 582	0.0247 0	
615	0.9386 741	0.4412 5	0.0002 760	675	0.0637 035	0.0232 4	
16	0.9233 452	0.4290 8	0.0002 586	76	0.0600 191	0.0218 8	
17	0.9071 833	0.4169 7	0.0002 421	77	0.0565 495	0.0206 0	
18	0.8902 642	0.4049 2	0.0002 264	78	0.0532 394	0.0193 8	
19	0.8726 405	0.3929 2	0.0002 116	79	0.0499 783	0.0181 8	

TABLE XII (*Continued*)

Wavelength	$\bar{x}$	$\bar{y}$	*Wavelength*	$\bar{x}$	$\bar{y}$
680 $m\mu$	0.0467 659	0.0170 0	725 $m\mu$	0.0020 386	0.0007 3625
81	0.0436 218	0.0158 47	26	0.0019 059	0.0006 8832
82	0.0406 766	0.0147 68	27	0.0017 818	0.0006 4351
83	0.0379 113	0.0137 56	28	0.0016 642	0.0006 0104
84	0.0353 183	0.0128 08	29	0.0015 509	0.0005 6013
685	0.0328 814	0.0119 18	730	0.0014 398	0.0005 2000
86	0.0305 959	0.0110 84	31	0.0013 342	0.0004 8184
87	0.0284 454	0.0103 00	32	0.0012 380	0.0004 4712
88	0.0264 137	0.0095 60	33	0.0011 504	0.0004 1548
89	0.0244 924	0.0088 61	34	0.0010 703	0.0003 8656
690	0.0226 734	0.0082 00	735	0.0009 968	0.0003 6000
91	0.0209 931	0.0075 90	36	0.0009 288	0.0003 3544
92	0.0194 933	0.0070 46	37	0.0008 653	0.0003 1252
93	0.0181 469	0.0065 58	38	0.0008 054	0.0002 9088
94	0.0169 406	0.0061 21	39	0.0007 480	0.0002 7016
695	0.0158 470	0.0057 25	740	0.0006 922	0.0002 5000
96	0.0148 467	0.0053 63	41	0.0006 397	0.0002 3102
97	0.0139 205	0.0050 28	42	0.0005 923	0.0002 1392
98	0.0130 409	0.0047 10	43	0.0005 496	0.0001 9850
99	0.0121 940	0.0044 04	44	0.0005 110	0.0001 8456
700	0.0113 525	0.0041 00	745	0.0004 759	0.0001 7188
1	0.0105 514	0.0038 1068	46	0.0004 437	0.0001 6024
2	0.0098 307	0.0035 5040	47	0.0004 138	0.0001 4944
3	0.0091 809	0.0033 1572	48	0.0003 857	0.0001 3928
4	0.0085 924	0.0031 0320	49	0.0003 587	0.0001 2954
705	0.0080 558	0.0029 0938	750	0.0003 323	0.0001 2000
6	0.0075 613	0.0027 3080	51	0.0003 074	0.0001 1103
7	0.0070 995	0.0025 6402	52	0.0002 853	0.0001 0304
8	0.0066 608	0.0024 0560	53	0.0002 656	0.0000 9591
9	0.0062 358	0.0022 5208	54	0.0002 479	0.0000 8952
710	0.0058 147	0.0021 0000	755	0.0002 319	0.0000 8375
11	0.0054 109	0.0019 5418	56	0.0002 173	0.0000 7848
12	0.0050 418	0.0018 2088	57	0.0002 038	0.0000 7359
13	0.0047 038	0.0016 9880	58	0.0001 909	0.0000 6896
14	0.0043 932	0.0015 8664	59	0.0001 785	0.0000 6447
715	0.0041 066	0.0014 8312	760	0.0001 661	0.0000 6000
16	0.0038 403	0.0013 8696	61	0.0001 543	0.0000 5572
17	0.0035 909	0.0012 9686	62	0.0001 435	0.0000 5184
18	0.0033 546	0.0012 1152	63	0.0001 338	0.0000 4832
19	0.0031 279	0.0011 2966	64	0.0001 249	0.0000 4512
720	0.0029 073	0.0010 5000	765	0.0001 168	0.0000 4219
21	0.0026 990	0.0009 7477	66	0.0001 093	0.0000 3948
22	0.0025 102	0.0009 0656	67	0.0001 023	0.0000 3695
23	0.0023 386	0.0008 4459	68	0.0000 957	0.0000 3456
24	0.0021 821	0.0007 8808	69	0.0000 893	0.0000 3226
			770	0.0000 831	0.0000 3000

TABLE XIII

Tristimulus Values for Spectrum Colors

(Weighted by Energy Distribution of Illuminant A)

Wavelength	$E_A\bar{x}$	$E_A\bar{y}$	$E_A\bar{z}$	*Wavelength*	$E_A\bar{x}$	$E_A\bar{y}$	$E_A\bar{z}$
380 $m\mu$	0.01 3	0.00 0	0.06 4	385 $m\mu$	0.02 5	0.00 1	0.11 6
81	0.01 6	0.00 0	0.07 4	86	0.02 8	0.00 1	0.13 2
82	0.01 8	0.00 1	0.08 3	87	0.03 2	0.00 1	0.15 1
83	0.02 0	0.00 1	0.09 3	88	0.03 7	0.00 1	0.17 5
84	0.02 2	0.00 1	0.10 4	89	0.04 3	0.00 1	0.20 5

TABLE XIII (*Continued*)

Wavelength	$E_A\bar{x}$	$E_A\bar{y}$	$E_A\bar{z}$	*Wavelength*	$E_A\bar{x}$	$E_A\bar{y}$	$E_A\bar{z}$
390 mμ	0.05 1	0.00 1	0.24 3	450 mμ	11.12 5	1.25 7	58.65 8
91	0.06 0	0.00 2	0.28 5	51	11.19 0	1.33 8	59.40 2
92	0.06 9	0.00 2	0.32 8	52	11.25 1	1.42 2	60.14 7
93	0.07 9	0.00 2	0.37 5	53	11.29 6	1.51 0	60.83 7
94	0.09.0	0.00 3	0.42 7	54	11.31 7	1.60 2	61.42 4
395	0.10 3	0.00 3	0.48 8	455	11.30 6	1.70 0	61.87 0
96	0.11 8	0.00 3	0.55 8	56	11.27 8	1.80 2	62.25 2
97	0.13 6	0.00 4	0.64 3	57	11.24 3	1.91 0	62.62 4
98	0.15 7	0.00 4	0.74 3	58	11.19 1	2.02 3	62.92 6
99	0.18 2	0.00 5	0.86 4	59	11.11 2	2.14 3	63.10 8
400	0.21 2	0.00 6	1.00 8	460	11.00 2	2.26 9	63.13 3
1	0.24 4	0.00 7	1.15 6	61	10.88 5	2.40 0	63.13 3
2	0.27 3	0.00 8	1.29 7	62	10.76 4	2.53 6	63.12 6
3	0.30 3	0.00 8	1.44 1	63	10.60 9	2.67 6	62.93 7
4	0.33 7	0.00 9	1.59 9	64	10.39 5	2.82 3	62.43 4
405	0.37 6	0.01 0	1.78 5	465	10.11 0	2.97 7	61.52 8
6	0.42 3	0.01 2	2.01 1	66	9.78 8	3.14 0	60.41 6
7	0.48 2	0.01 3	2.29 3	67	9.46 0	3.31 2	59.27 7
8	0.55 6	0.01 5	2.64 6	68	9.11 7	3.49 6	58.05 1
9	0.64 8	0.01 8	3.08 7	69	8.75 6	3.69 2	56.70 2
410	0.76 3	0.02 1	3.63 6	470	8.37 6	3.90 1	55.21 7
11	0.89 1	0.02 5	4.25 1	71	7.99 7	4.12 3	53.72 4
12	1.02 6	0.02 8	4.89 5	72	7.62 7	4.35 6	52.27 2
13	1.17 0	0.03 2	5.58 9	73	7.25 4	4.60 0	50.77 5
14	1.32 9	0.03 7	6.35 0	74	6.87 0	4.85 7	49.18 0
415	1.50 4	0.04 2	7.19 5	475	6.47 3	5.12 7	47.46 3
16	1.70 2	0.04 8	8.14 8	76	6.08 1	5.41 1	45.73 8
17	1.93 0	0.05 5	9.24 3	77	5.70 3	5.71 0	44.08 6
18	2.19 0	0.06 3	10.50 1	78	5.33 5	6.02 5	42.46 5
19	2.48 7	0.07 3	11.93 7	79	4.97 2	6.35 7	40.84 8
420	2.82 3	0.08 4	13.56 4	480	4.61 4	6.70 7	39.22 0
21	3.19 3	0.09 7	15.35 8	81	4.26 0	7.06 9	37.60 2
22	3.58 6	0.11 2	17.27 5	82	3.91 3	7.43 9	36.01 7
23	3.99 9	0.12 8	19.28 9	83	3.57 7	7.82 2	34.46 6
24	4.42 5	0.14 6	21.37 6	84	3.25 8	8.22 2	32.94 6
425	4.86 0	0.16 6	23.51 1	485	2.95 8	8.64 2	31.45 8
26	5.30 1	0.18 7	25.68 6	86	2.67 6	9.08 7	30.01 6
27	5.74 5	0.20 9	27.88 1	87	2.41 1	9.56 2	28.64 7
28	6.18 3	0.23 3	30.05 6	88	2.16 4	10.07 2	27.36 4
29	6.60 6	0.25 9	32.17 1	89	1.93 6	10.62 0	26.17 3
430	7.00 5	0.28 6	34.18 3	490	1.72 6	11.21 3	25.07 2
31	7.38 1	0.31 5	36.09 7	91	1.52 6	11.84 4	24.00 2
32	7.74 0	0.34 6	37.93 9	92	1.33 2	12.50 3	22.93 8
33	8.08 0	0.37 8	39.70 5	93	1.15 0	13.19 7	21.91 6
34	8.40 2	0.41 2	41.39 4	94	0.98 4	13.92 7	20.95 8
435	8.70 5	0.44 9	43.00 4	495	0.83 7	14.69 9	20.08 1
36	8.99 0	0.48 7	44.53 6	96	0.70 4	15.51 7	19.24 9
37	9.25 8	0.52 7	45.99 3	97	0.58 0	16.38 5	18.43 5
38	9.50 8	0.56 9	47.37 7	98	0.46 8	17.30 7	17.66 1
39	9.74 0	0.61 4	48.69 3	99	0.37 2	18.28 9	16.94 0
440	9.95 6	0.66 0	49.94 4	500	0.29 4	19.33 5	16.28 2
41	10.15 6	0.70 8	51.13 2	1	0.23 3	20.45 3	15.66 8
42	10.33 8	0.75 8	52.24 9	2	0.18 5	21.64 8	15.06 9
43	10.49 8	0.81 0	53.28 0	3	0.15 3	22.91 2	14.48 4
44	10.63 5	0.86 4	54.21 2	4	0.14 0	24.24 0	13.91 3
445	10.74 6	0.92 0	55.03 6	505	0.15 3	25.62 8	13.35 8
46	10.83 6	0.98 0	55.78 6	6	0.19 0	27.06 9	12.79 8
47	10.91 7	1.04 3	56.51 4	7	0.25 2	28.55 5	12.21 5
48	10.99 1	1.11 0	57.22 8	8	0.34 2	30.08 2	11.62 1
49	11.06 0	1.18 1	57.94 0	9	0.46 2	31.64 3	11.02 8

TABLE XIII (*Continued*)

Wavelength	$E_A\bar{x}$	$E_A\bar{y}$	$E_A\bar{z}$	*Wavelength*	$E_A\bar{x}$	$E_A\bar{y}$	$E_A\bar{z}$
510 $m\mu$	0.61 6	33.22 8	10.44 9	570 $m\mu$	81.69 1	102.03 5	0.22 1
11	0.81 0	34.87 1	9.88 1	71	84.02 1	102.02 4	0.21 3
12	1.04 8	36.59 7	9.31 8	72	86.35 5	101.95 5	0.20 6
13	1.33 0	38.39 0	8.76 6	73	88.69 1	101.83 2	0.20 0
14	1.65 4	40.23 8	8.23 4	74	91.02 3	101.65 4	0.19 6
515	2.01 8	42.12 1	7.73 4	575	93.34 9	101.42 4	0.19 1
16	2.43 1	44.02 4	7.27 3	76	95.66 7	101.14 4	0.18 7
17	2.90 0	45.93 1	6.84 2	77	97.98 4	100.81 7	0.18 4
18	3.42 1	47.82 0	6.43 5	78	100.29 0	100.44 3	0.18 1
19	3.98 6	49.67 5	6.04 7	79	102.58 3	100.02 5	0.17 9
520	4.58 7	51.47 5	5.67 3	580	104.85 4	99.56 3	0.17 6
21	5.23 1	53.23 5	5.33 7	81	107.09 2	99.04 7	0.17 3
22	5.93 1	54.98 0	5.05 3	82	109.28 3	98.46 4	0.16 9
23	6.68 2	56.71 1	4.80 5	83	111.42 5	97.82 1	0.16 6
24	7.47 8	58.42 3	4.57 3	84	113.51 5	97.12 7	0.16 3
525	8.30 9	60.11 6	4.33 7	585	115.54 5	96.38 3	0.15 9
26	9.18 1	61.78 7	4.10 5	86	117.52 5	95.59 6	0.15 6
27	10.10 3	63.43 4	3.89 9	87	119.46 5	94.77 5	0.15 2
28	11.06 8	65.05 6	3.70 8	88	121.35 4	93.92 2	0.14 9
29	12.06 8	66.64 9	3.52 3	89	123.18 8	93.04 4	0.14 5
530	13.09 4	68.21 0	3.33 5	590	124.96 6	92.15 0	0.14 1
31	14.14 3	69.73 8	3.14 6	91	126.67 3	91.22 6	0.13 7
32	15.22 0	71.23 0	2.96 6	92	128.28 9	90.25 9	0.13 3
33	16.32 7	72.68 7	2.79 4	93	129.80 6	89.25 4	0.12 9
34	17.46 4	74.11 1	2.62 7	94	131.20 7	88.21 2	0.12 5
535	18.63 4	75.50 1	2.46 3	595	132.48 2	87.13 9	0.12 1
36	19.83 3	76.85 9	2.30 5	96	133.62 2	86.03 6	0.11 6
37	21.06 3	78.18 7	2.15 6	97	134.63 6	84.90 8	0.11 2
38	22.32 5	79.48 4	2.01 6	98	135.53 4	83.76 1	0.10 7
39	23.62 3	80.75 5	1.88 1	99	136.33 0	82.59 8	0.10 3
540	24.95 9	81.99 6	1.75 0	600	137.03 7	81.42 4	0.09 8
41	26.33 1	83.20 4	1.62 5	1	137.64 3	80.23 1	0.09 4
42	27.73 8	84.37 1	1.50 8	2	138.12 6	79.01 1	0.08 9
43	29.17 8	85.49 9	1.39 9	3	138.47 1	77.76 6	0.08 4
44	30.65 3	86.58 9	1.29 6	4	138.67 0	76.49 7	0.08 0
545	32.16 3	87.64 4	1.19 8	605	138.71 0	75.20 8	0.07 5
46	33.70 9	88.66 5	1.10 7	6	138.59 1	73.90 1	0.07 0
47	35.29 1	89.65 4	1.02 5	7	138.32 5	72.58 2	0.06 5
48	36.91 2	90.61 3	0.95 0	8	137.92 0	71.25 3	0.06 0
49	38.57 2	91.54 3	0.88 1	9	137.38 1	69.91 6	0.05 5
550	40.27 4	92.44 5	0.81 5	610	136.72 7	68.57 9	0.05 1
51	42.01 6	93.31 9	0.75 2	11	135.94 2	67.23 7	0.04 8
52	43.79 9	94.15 9	0.69 7	12	135.01 4	65.88 5	0.04 5
53	45.62 0	94.96 4	0.64 7	13	133.94 5	64.52 4	0.04 3
54	47.48 1	95.73 3	0.60 1	14	132.74 1	63.15 2	0.04 1
555	49.38 1	96.46 4	0.55 6	615	131.40 5	61.77 1	0.03 9
56	51.32 1	97.15 6	0.51 5	16	129.93 1	60.37 9	0.03 6
57	53.30 1	97.80 7	0.47 8	17	128.31 6	58.97 8	0.03 4
58	55.31 8	98.41 6	0.44 5	18	126.56 9	57.56 8	0.03 2
59	57.36 8	98.98 1	0.41 5	19	124.69 6	56.14 7	0.03 0
560	59.44 7	99.50 0	0.38 6	620	122.71 6	54.71 9	0.02 8
61	61.55 4	99.97 1	0.35 9	21	120.58 6	53.26 9	0.02 7
62	63.69 0	100.39 3	0.33 5	22	118.28 6	51.78 9	0.02 5
63	65.85 2	100.76 6	0.31 5	23	115.84 3	50.28 8	0.02 3
64	68.04 1	101.09 0	0.29 6	24	113.28 4	48.77 6	0.02 2
565	70.25 6	101.36 6	0.27 8	625	110.63 9	47.26 1	0.02 1
66	72.49 7	101.59 4	0.26 3	26	107.92 4	45.75 2	0.01 9
67	74.76 7	101.77 6	0.25 0	27	105.16 0	44.25 9	0.01 8
68	77.05 8	101.90 8	0.23 9	28	102.37 9	42.79 2	0.01 6
69	79.36 9	101.99 6	0.23 0	29	99.61 4	41.35 9	0.01 5

TABLE XIII (*Continued*)

Wavelength	$E_A\bar{x}$	$E_A\bar{y}$	$E_A\bar{z}$	*Wavelength*	$E_A\bar{x}$	$E_A\bar{y}$
630 mμ	96.90 0	39.97 0	0.01 4	690 mμ	4.35 2	1.57 4
31	94.22 9	38.62 2	0.01 2	91	4.04 3	1.46 2
32	91.56 8	37.30 3	0.01 1	92	3.76 6	1.36 1
33	88.92 6	36.01 3	0.01 0	93	3.51 8	1.27 1
34	86.29 3	34.74 8	0.00 9	94	3.29 5	1.19 0
635	83.67 6	33.50 9	0.00 8	695	3.09 2	1.11 7
36	81.07 2	32.29 4	0.00 7	96	2.90 6	1.05 0
37	78.47 5	31.10 1	0.00 7	97	2.73 4	0.98 7
38	75.89 5	29.93 1	0.00 6	98	2.56 9	0.92 8
39	73.31 7	28.77 9	0.00 5	99	2.41 0	0.87 0
640	70.75 2	27.64 7	0.00 5	700	2.25 1	0.81 3
41	68.20 1	26.53 4	0.00 4	1	2.09 8	0.75 8
42	65.67 6	25.44 6	0.00 4	2	1.96 1	0.70 8
43	63.18 0	24.38 2	0.00 3	3	1.83 7	0.66 4
44	60.71 5	23.34 2	0.00 3	4	1.72 5	0.62 3
645	58.28 7	22.32 9	0.00 2	705	1.62 2	0.58 6
46	55.89 6	21.34 1	0.00 2	6	1.52 7	0.55 2
47	53.54 6	20.38 0	0.00 2	7	1.43 8	0.51 9
48	51.24 2	19.44 5	0.00 1	8	1.35 3	0.48 9
49	48.98 8	18.53 8	0.00 1	9	1.27 1	0.45 9
650	46.79 0	17.65 8	0.00 1	710	1.18 9	0.42 9
51	44.66 1	16.81 2		11	1.10 9	0.40 1
52	42.59 0	15.99 7		12	1.03 7	0.37 4
53	40.61 5	15.21 6		13	0.97 0	0.35 0
54	38.69 6	14.46 6		14	0.90 9	0.32 8
655	36.84 0	13.74 4		715	0.85 2	0.30 8
56	35.04 7	13.04 9		16	0.79 9	0.28 8
57	33.30 7	12.37 8		17	0.74 9	0.27 1
58	31.61 5	11.72 9		18	0.70 2	0.25 3
59	29.97 2	11.10 2		19	0.65 6	0.23 7
660	28.36 1	10.49 0		720	0.61 2	0.22 1
61	26.80 6	9.90 1		21	0.57 0	0.20 6
62	25.31 8	9.34 0		22	0.53 1	0.19 2
63	23.89 2	8.80 5		23	0.49 6	0.17 9
64	22.53 9	8.29 8		24	0.46 4	0.16 8
665	21.24 8	7.81 5		725	0.43 5	0.15 7
66	20.01 9	7.35 6		26	0.40 8	0.14 7
67	18.84 5	6.91 8		27	0.38 2	0.13 8
68	17.72 0	6.49 9		28	0.35 8	0.12 9
69	16.65 1	6.10 2		29	0.33 4	0.12 1
670	15.62 5	5.72 1		730	0.31 1	0.11 2
71	14.66 6	5.36 5		31	0.28 9	0.10 4
72	13.79 5	5.04 3		32	0.26 9	0.09 7
73	13.00 5	4.75 1		33	0.25 1	0.09 0
74	12.27 7	4.48 2		34	0.23 4	0.08 4
675	11.60 2	4.23 2		735	0.21 8	0.07 9
76	10.97 1	3.99 9		36	0.20 4	0.07 4
77	10.37 4	3.77 9		37	0.19 0	0.06 9
78	9.80 2	3.56 8		38	0.17 8	0.06 4
79	9.23 5	3.35 9		39	0.16 5	0.06 0
680	8.67 2	3.15 2		740	0.15 3	0.05 5
81	8.11 7	2.94 9		41	0.14 2	0.05 1
82	7.59 6	2.75 8		42	0.13 2	0.04 8
83	7.10 4	2.57 8		43	0.12 3	0.04 4
84	6.64 2	2.40 9		44	0.11 4	0.04 1
685	6.20 5	2.24 9		745	0.10 7	0.03 9
86	5.79 3	2.09 9		46	0.10 0	0.03 6
87	5.40 5	1.95 7		47	0.09 3	0.03 4
88	5.03 6	1.82 3		48	0.08 7	0.03 1
89	4.68 5	1.69 5		49	0.08 1	0.02 9

TABLE XIII (*Continued*)

Wavelength	$E_A\bar{x}$	$E_A\bar{y}$	*Wavelength*	$E_A\bar{x}$	$E_A\bar{y}$
750 mμ	0.07 5	0.02 7	761 mμ	0.03 6	0.01 3
51	0.07 0	0.02 5	62	0.03 3	0.01 2
52	0.06 5	0.02 3	63	0.03 1	0.01 1
53	0.06 1	0.02 2	64	0.02 9	0.01 1
54	0.05 7	0.02 1	765	0.02 7	0.01 0
755	0.05 3	0.01 9	66	0.02 6	0.00 9
56	0.05 0	0.01 8	67	0.02 4	0.00 9
57	0.04 7	0.01 7	68	0.02 3	0.00 8
58	0.04 4	0.01 6	69	0.02 1	0.00 8
59	0.04 1	0.01 5	770	0.02 0	0.00 7
760	0.03 9	0.01 4			

TABLE XIV

TRISTIMULUS VALUES FOR SPECTRUM COLORS

(Weighted by Energy Distribution of Illuminant B)

Wavelength	$E_B\bar{x}$	$E_B\bar{y}$	$E_B\bar{z}$	*Wavelength*	$E_B\bar{x}$	$E_B\bar{y}$	$E_B\bar{z}$
380 mμ	0.03 1	0.00 1	0.14 7	420 mμ	8.49 6	0.25 3	40.82 0
81	0.03 6	0.00 1	0.17 2	21	9.60 9	0.29 2	46.22 3
82	0.04 2	0.00 1	0.19 7	22	10.79 2	0.33 6	51.98 0
83	0.04 7	0.00 1	0.22 2	23	12.02 6	0.38 5	58.00 6
84	0.05 3	0.00 2	0.25 2	24	13.29 4	0.43 9	64.21 8
385	0.06 1	0.00 2	0.28 6	425	14.57 9	0.49 7	70.53 4
86	0.06 9	0.00 2	0.32 8	26	15.87 5	0.55 9	76.92 0
87	0.08 0	0.00 2	0.38 0	27	17.16 7	0.62 5	83.31 7
88	0.09 4	0.00 3	0.44 5	28	18.43 0	0.69 6	89.59 6
89	0.11 2	0.00 3	0.52 7	29	19.63 6	0.77 0	95.62 9
390	0.13 3	0.00 4	0.63 0	430	20.75 7	0.84 8	101.28 8
91	0.15 8	0.00 4	0.74 5	31	21.79 5	0.93 0	106.58 6
92	0.18 3	0.00 5	0.86 7	32	22.76 8	1.01 6	111.60 3
93	0.21 1	0.00 6	0.99 9	33	23.67 1	1.10 7	116.31 8
94	0.24 3	0.00 7	1.14 8	34	24.50 3	1.20 2	120.72 0
395	0.27 9	0.00 8	1.32 0	435	25.26 2	1.30 2	124.79 7
96	0.32 2	0.00 9	1.52 4	36	25.95 1	1.40 5	128.55 7
97	0.37 3	0.01 0	1.76 8	37	26.57 2	1.51 3	132.01 2
98	0.43 5	0.01 2	2.05 9	38	27.12 5	1.62 4	135.16 7
99	0.50 8	0.01 4	2.40 9	39	27.61 1	1.73 9	138.03 0
400	0.59 6	0.01 7	2.82 9	440	28.02 9	1.85 8	140.60 8
1	0.68 8	0.01 9	3.26 7	41	28.38 3	1.98 0	142.90 6
2	0.77 7	0.02 1	3.68 6	42	28.67 0	2.10 3	144.90 4
3	0.86 7	0.02 4	4.11 7	43	28.88 2	2.22 8	146.57 5
4	0.96 7	0.02 7	4.59 4	44	29.01 4	2.35 7	147.89 6
405	1.08 4	0.03 0	5.15 2	445	29.06 4	2.49 0	148.85 5
6	1.22 7	0.03 4	5.83 3	46	29.05 0	2.62 7	149.54 9
7	1.40 4	0.03 9	6.67 9	47	28.99 9	2.77 1	150.11 1
8	1.62 6	0.04 5	7.73 9	48	28.92 1	2.92 1	150.58 1
9	1.90 3	0.05 3	9.06 5	49	28.82 3	3.07 9	150.99 6
410	2.24 8	0.06 3	10.71 4	450	28.71 2	3.24 5	151.38 8
11	2.63 6	0.07 3	12.57 2	51	28.59 6	3.41 8	151.79 8
12	3.04 4	0.08 5	14.52 4	52	28.46 2	3.59 7	152.15 4
13	3.48 3	0.09 7	16.63 3	53	28.28 6	3.78 1	152.33 3
14	3.96 5	0.11 0	18.95 1	54	28.04 9	3.97 2	152.23 9
415	4.50 0	0.12 5	21.52 0	455	27.74 0	4.17 0	151.80 0
16	5.10 2	0.14 3	24.41 9	56	27.39 4	4.37 7	151.20 5
17	5.79 2	0.16 4	27.74 6	57	27.03 4	4.59 2	150.57 5
18	6.58 2	0.18 9	31.56 0	58	26.63 8	4.81 6	149.78 8
19	7.48 2	0.21 8	35.90 8	59	26.19 2	5.05 1	148.75 0

TABLE XIV (*Continued*)

Wavelength	$E_B\bar{x}$	$E_B\bar{y}$	$E_B\bar{z}$	*Wavelength*	$E_B\bar{x}$	$E_B\bar{y}$	$E_B\bar{z}$
460 $m\mu$	25.68 7	5.29 8	147.39 9	520 $m\mu$	5.66 2	63.54 5	7.00 4
61	25.18 3	5.55 4	146.05 9	21	6.40 7	65.20 4	6.53 7
62	24.68 4	5.81 5	144.75 6	22	7.21 2	66.85 6	6.14 5
63	24.11 8	6.08 4	143.08 2	23	8.07 1	68.49 7	5.80 3
64	23.43 3	6.36 4	140.73 7	24	8.97 5	70.12 2	5.48 9
465	22.59 9	6.65 5	137.53 0	525	9.91 4	71.72 8	5.17 4
66	21.69 8	6.96 0	133.92 7	26	10.89 5	73.31 8	4.87 2
67	20.80 0	7.28 3	130.33 4	27	11.92 8	74.89 6	4.60 3
68	19.88 5	7.62 5	126.61 1	28	13.00 7	76.45 3	4.35 7
69	18.94 4	7.98 7	122.67 7	29	14.12 0	77.98 3	4.12 3
470	17.97 5	8.37 2	118.49 7	530	15.25 6	79.47 6	3.88 6
71	17.02 3	8.77 6	114.35 4	31	16.41 3	80.93 5	3.65 1
72	16.10 3	9.19 6	110.35 9	32	17.59 9	82.36 4	3.43 0
73	15.19 0	9.63 3	106.32 4	33	18.81 4	83.75 8	3.22 0
74	14.26 7	10.08 7	102.13 8	34	20.05 7	85.11 3	3.01 7
475	13.33 2	10.55 9	97.75 2	535	21.33 0	86.42 5	2.82 0
76	12.41 8	11.05 0	93.40 8	36	22.63 1	87.70 0	2.63 0
77	11.54 9	11.56 3	89.27 5	37	23.96 2	88.94 6	2.45 3
78	10.71 1	12.09 7	85.25 9	38	25.32 3	90.15 5	2.28 6
79	9.89 7	12.65 3	81.30 5	39	26.71 6	91.32 4	2.12 6
480	9.10 4	13.23 3	77.38 2	540	28.13 8	92.44 3	1.97 3
81	8.33 2	13.82 4	73.53 8	41	29.59 3	93.51 0	1.82 6
82	7.58 5	14.42 1	69.81 9	42	31.07 5	94.52 2	1.69 0
83	6.87 2	15.02 7	66.21 3	43	32.58 4	95.47 8	1.56 2
84	6.20 1	15.64 9	62.71 2	44	34.11 7	96.37 2	1.44 3
485	5.57 7	16.29 3	59.31 1	545	35.67 1	97.20 2	1.32 9
86	4.99 7	16.96 5	56.03 8	46	37.24 7	97.97 2	1.22 3
87	4.45 7	17.67 4	52.94 9	47	38.84 8	98.69 0	1.12 8
88	3.95 9	18.42 5	50.05 9	48	40.47 1	99.35 1	1.04 2
89	3.50 4	19.22 2	47.37 2	49	42.11 6	99.95 4	0.96 2
490	3.08 9	20.07 2	44.88 0	550	43.78 1	100.49 5	0.88 5
91	2.70 0	20.95 7	42.47 1	51	45.46 2	100.97 2	0.81 4
92	2.32 9	21.86 0	40.10 4	52	47.15 9	101.38 3	0.75 1
93	1.98 6	22.78 9	37.84 6	53	48.87 0	101.73 0	0.69 3
94	1.67 8	23.74 9	35.73 9	54	50.59 5	102.01 0	0.64 0
495	1.40 9	24.74 7	33.80 8	555	52.33 1	102.22 6	0.59 0
96	1.16 9	25.78 5	31.98 5	56	54.07 7	102.37 3	0.54 3
97	0.95 1	26.86 5	30.22 7	57	55.83 1	102.45 0	0.50 1
98	0.75 7	27.99 4	28.56 6	58	57.59 1	102.46 0	0.46 3
99	0.59 4	29.17 9	27.02 6	59	59.35 2	102.40 5	0.42 9
500	0.46 3	30.42 7	25.62 3	560	61.11 1	102.28 6	0.39 6
1	0.36 2	31.74 4	24.31 7	61	62.86 7	102.10 3	0.36 7
2	0.28 2	33.13 1	23.06 3	62	64.61 5	101.85 2	0.34 0
3	0.23 0	34.57 8	21.85 8	63	66.35 6	101.53 6	0.31 7
4	0.20 9	36.07 4	20.70 5	64	68.08 7	101.15 8	0.29 6
505	0.22 4	37.61 8	19.60 7	565	69.80 8	100.72 1	0.27 7
6	0.27 5	39.19 3	18.53 0	66	71.51 8	100.22 2	0.25 9
7	0.36 0	40.78 5	17.44 7	67	73.21 8	99.66 7	0.24 5
8	0.48 2	42.39 1	16.37 7	68	74.90 0	99.05 4	0.23 3
9	0.64 3	44.00 5	15.33 7	69	76.56 4	98.39 0	0.22 2
510	0.84 8	45.62 2	14.34 6	570	78.20 0	97.67 5	0.21 1
11	1.09 8	47.28 5	13.39 9	71	79.80 4	96.90 4	0.20 2
12	1.40 4	49.02 4	12.48 3	72	81.37 2	96.07 1	0.19 4
13	1.76 1	50.82 2	11.60 5	73	82.90 3	95.18 6	0.18 7
14	2.16 5	52.66 3	10.77 7	74	84.39 5	94.25 1	0.18 1
515	2.61 2	54.53 0	10.01 3	575	85.85 0	93.27 7	0.17 6
16	3.11 4	56.40 1	9.31 8	76	87.26 7	92.26 2	0.17 1
17	3.67 9	58.25 6	8.67 8	77	88.64 5	91.20 8	0.16 6
18	4.29 8	60.07 5	8.08 4	78	89.98 4	90.12 1	0.16 3
19	4.96 3	61.84 3	7.52 8	79	91.28 2	89.00 6	0.15 9

TABLE XIV (*Continued*)

Wavelength	$E_B\bar{x}$	$E_B\bar{y}$	$E_B\bar{z}$	*Wavelength*	$E_B\bar{x}$	$E_B\bar{y}$	$E_B\bar{z}$
580 mμ	92.53 9	87.87 0	0.15 5	640 mμ	45.77 1	17.88 5	0.00 3
81	93.74 6	86.70 3	0.15 1	41	43.99 1	17.11 5	0.00 3
82	94.88 6	85.49 2	0.14 7	42	42.24 4	16.36 7	0.00 2
83	95.96 2	84.24 6	0.14 3	43	40.52 8	15.64 0	0.00 2
84	96.97 5	82.97 5	0.13 9	44	38.84 3	14.93 4	0.00 2
585	97.92 1	81.68 2	0.13 5	645	37.18 9	14.24 7	0.00 2
86	98.81 1	80.37 4	0.13 1	46	35.56 8	13.58 0	0.00 1
87	99.65 1	79.05 6	0.12 7	47	33.98 3	12.93 4	0.00 1
88	100.43 5	77.73 2	0.12 3	48	32.43 5	12.30 8	0.00 1
89	101.16 4	76.40 8	0.11 9	49	30.92 6	11.70 3	0.00 1
590	101.83 7	75.09 4	0.11 5	650	29.45 8	11.11 7	
91	102.44 2	73.77 6	0.11 1	51	28.04 0	10.55 5	
92	102.96 2	72.44 0	0.10 7	52	26.67 2	10.01 6	
93	103.39 9	71.09 7	0.10 3	53	25.35 8	9.50 0	
94	103.74 8	69.75 1	0.09 9	54	24.09 0	9.00 6	
595	104.00 8	68.41 0	0.09 5	655	22.86 6	8.53 0	
96	104.17 1	67.07 3	0.09 1	56	21.68 5	8.07 4	
97	104.24 4	65.74 2	0.08 7	57	20.54 4	7.63 5	
98	104.24 1	64.42 2	0.08 3	58	19.43 7	7.21 1	
99	104.17 8	63.11 8	0.07 9	59	18.36 5	6.80 2	
600	104.07 3	61.83 8	0.07 5	660	17.31 7	6.40 5	
1	103.92 0	60.57 4	0.07 1	61	16.30 9	6.02 4	
2	103.70 0	59.31 9	0.06 7	62	15.34 7	5.66 2	
3	103.40 1	58.07 0	0.06 3	63	14.42 7	5.31 7	
4	103.01 1	56.82 6	0.05 9	64	13.55 8	4.99 2	
605	102.52 2	55.58 7	0.05 5	665	12.73 1	4.68 2	
6	101.93 7	54.35 6	0.05 1	66	11.94 6	4.39 0	
7	101.26 8	53.13 7	0.04 8	67	11.19 9	4.11 1	
8	100.51 8	51.93 0	0.04 4	68	10.48 7	3.84 6	
9	99.68 5	50.73 2	0.04 0	69	9.81 3	3.59 6	
610	98.77 9	49.54 6	0.03 7	670	9.16 9	3.35 7	
11	97.79 1	48.36 7	0.03 5	71	8.56 9	3.13 5	
12	96.71 4	47.19 5	0.03 2	72	8.02 6	2.93 4	
13	95.55 2	46.02 9	0.03 1	73	7.53 3	2.75 2	
14	94.30 8	44.86 8	0.02 9	74	7.08 0	2.58 5	
615	92.98 5	43.71 0	0.02 7	675	6.66 0	2.43 0	
16	91.58 0	42.55 7	0.02 6	76	6.26 9	2.28 5	
17	90.09 2	41.40 9	0.02 4	77	5.90 0	2.14 9	
18	88.52 7	40.26 5	0.02 3	78	5.54 8	2.01 9	
19	86.88 8	39.12 3	0.02 1	79	5.20 1	1.89 2	
620	85.18 9	37.98 6	0.02 0	680	4.85 9	1.76 6	
21	83.40 0	36.84 2	0.01 8	81	4.52 4	1.64 4	
22	81.50 9	35.68 7	0.01 7	82	4.21 1	1.52 9	
23	79.53 7	34.52 7	0.01 6	83	3.91 6	1.42 1	
24	77.50 0	33.36 8	0.01 5	84	3.64 0	1.32 0	
625	75.41 8	32.21 6	0.01 4	685	3.38 2	1.22 6	
26	73.30 4	31.07 6	0.01 3	86	3.13 9	1.13 7	
27	71.17 3	29.95 5	0.01 2	87	2.91 2	1.05 4	
28	69.04 5	28.85 9	0.01 1	88	2.69 7	0.97 6	
29	66.94 3	27.79 4	0.01 0	89	2.49 5	0.90 2	
630	64.88 7	26.76 5	0.00 9	690	2.30 4	0.83 3	
31	62.87 1	25.76 9	0.00 8	91	2.12 8	0.76 9	
32	60.87 3	24.79 8	0.00 7	92	1.97 1	0.71 2	
33	58.90 2	23.85 4	0.00 7	93	1.83 1	0.66 2	
34	56.95 3	22.93 3	0.00 6	94	1.70 5	0.61 6	
635	55.03 3	22.03 9	0.00 5	695	1.59 2	0.57 5	
36	53.13 7	21.16 7	0.00 5	96	1.48 7	0.53 7	
37	51.26 0	20.31 5	0.00 4	97	1.39 0	0.50 2	
38	49.41 0	19.48 6	0.00 4	98	1.29 9	0.46 9	
39	47.57 8	18.67 5	0.00 4	99	1.21 2	0.43 8	

TABLE XIV (*Continued*)

Wavelength	$E_B\bar{x}$	$E_B\bar{y}$	*Wavelength*	$E_B\bar{x}$	$E_B\bar{y}$
700 mμ	1.12 5	0.40 6	735 mμ	0.08 8	0.03 2
1	1.04 3	0.37 7	36	0.08 2	0.02 9
2	0.96 9	0.35 0	37	0.07 6	0.02 7
3	0.90 2	0.32 6	38	0.07 0	0.02 5
4	0.84 2	0.30 4	39	0.06 5	0.02 4
705	0.78 7	0.28 4	740	0.06 0	0.02 2
6	0.73 7	0.26 6	41	0.05 5	0.02 0
7	0.68 9	0.24 9	42	0.05 1	0.01 8
8	0.64 5	0.23 3	43	0.04 7	0.01 7
9	0.60 2	0.21 7	44	0.04 4	0.01 6
710	0.55 9	0.20 2	745	0.04 1	0.01 5
11	0.51 9	0.18 7	46	0.03 8	0.01 4
12	0.48 2	0.17 4	47	0.03 5	0.01 3
13	0.44 8	0.16 2	48	0.03 3	0.01 2
14	0.41 7	0.15 1	49	0.03 1	0.01 1
715	0.38 8	0.14 0	750	0.02 8	0.01 0
16	0.36 2	0.13 1	51	0.02 6	0.00 9
17	0.33 7	0.12 2	52	0.02 4	0.00 9
18	0.31 4	0.11 3	53	0.02 2	0.00 8
19	0.29 2	0.10 5	54	0.02 1	0.00 8
720	0.27 0	0.09 8	755	0.02 0	0.00 7
21	0.25 0	0.09 0	56	0.01 8	0.00 7
22	0.23 1	0.08 4	57	0.01 7	0.00 6
23	0.21 5	0.07 8	58	0.01 6	0.00 6
24	0.20 0	0.07 2	59	0.01 5	0.00 5
725	0.18 6	0.06 7	760	0.01 4	0.00 5
26	0.17 3	0.06 2	61	0.01 3	0.00 5
27	0.16 1	0.05 8	62	0.01 2	0.00 4
28	0.15 0	0.05 4	63	0.01 1	0.00 4
29	0.13 9	0.05 0	64	0.01 1	0.00 4
730	0.12 9	0.04 6	765	0.01 0	0.00 4
31	0.11 9	0.04 3	66	0.00 9	0.00 3
32	0.11 0	0.04 0	67	0.00 9	0.00 3
33	0.10 2	0.03 7	68	0.00 8	0.00 3
34	0.09 4	0.03 4	69	0.00 8	0.00 3

TABLE XV

TRISTIMULUS VALUES FOR SPECTRUM COLORS

(Weighted by Energy Distribution of Illuminant C)

Wavelength	$E_C\bar{x}$	$E_C\bar{y}$	$E_C\bar{z}$	*Wavelength*	$E_C\bar{x}$	$E_C\bar{y}$	$E_C\bar{z}$
380 mμ	0.04 6	0.00 1	0.21 6	400 mμ	0.91 4	0.02 5	4.33 5
81	0.05 4	0.00 2	0.25 4	1	1.05 6	0.02 9	5.01 1
82	0.06 2	0.00 2	0.29 1	2	1.19 3	0.03 3	5.66 0
83	0.07 0	0.00 2	0.33 0	3	1.33 3	0.03 7	6.32 9
84	0.07 9	0.00 2	0.37 3	4	1.48 8	0.04 1	7.06 8
385	0.09 0	0.00 3	0.42 5	405	1.67 0	0.04 6	7.93 6
86	0.10 4	0.00 3	0.48 8	6	1.89 1	0.05 2	8.99 1
87	0.12 0	0.00 3	0.56 8	7	2.16 6	0.06 0	10.30 6
88	0.14 2	0.00 4	0.66 8	8	2.51 1	0.07 0	11.95 2
89	0.16 8	0.00 5	0.79 5	9	2.94 2	0.08 2	14.01 3
390	0.20 2	0.00 6	0.95 3	410	3.47 8	0.09 7	16.57 5
91	0.23 9	0.00 7	1.13 1	11	4.08 1	0.11 3	19.46 2
92	0.27 8	0.00 8	1.31 7	12	4.71 5	0.13 1	22.50 1
93	0.32 1	0.00 9	1.51 9	13	5.39 9	0.14 9	25.78 4
94	0.36 9	0.01 0	1.74 7	14	6.15 0	0.17 0	29.39 3
395	0.42 5	0.01 2	2.01 3	415	6.98 2	0.19 4	33.39 1
96	0.49 1	0.01 4	2.32 7	16	7.91 8	0.22 2	37.89 9
97	0.57 0	0.01 6	2.70 1	17	8.99 1	0.25 5	43.07 2
98	0.66 5	0.01 8	3.14 9	18	10.21 9	0.29 4	48.99 9
99	0.77 8	0.02 2	3.68 9	19	11.61 6	0.33 8	55.74 8

TABLE XV (*Continued*)

Wavelength	$E_C\bar{x}$	$E_C\bar{y}$	$E_C\bar{z}$	*Wavelength*	$E_C\bar{x}$	$E_C\bar{y}$	$E_C\bar{z}$
420 mμ	13.18 8	0.39 2	63.36 1	480 mμ	11.84 9	17.22 2	100.71 1
21	14.91 0	0.45 3	71.72 3	81	10.80 5	17.92 8	95.36 5
22	16.73 7	0.52 1	80.61 6	82	9.80 0	18.63 3	90.21 1
23	18.64 0	0.59 7	89.90 8	83	8.84 7	19.34 4	85.23 3
24	20.59 1	0.67 9	99.46 3	84	7.95 2	20.06 8	80.41 9
425	22.56 1	0.76 8	109.14 8	485	7.12 4	20.81 2	75.76 1
26	24.54 0	0.86 3	118.90 9	86	6.35 7	21.58 4	71.29 4
27	26.50 8	0.96 5	128.64 9	87	5.64 7	22.39 4	67.08 8
28	28.42 2	1.07 2	138.16 9	88	4.99 5	23.24 7	63.16 0
29	30.23 9	1.18 5	147.26 7	89	4.40 2	24.14 9	59.51 3
430	31.91 6	1.30 4	155.74 2	490	3.86 4	25.10 6	56.13 4
31	33.45 7	1.42 7	163.61 8	91	3.36 1	26.09 4	52.88 2
32	34.88 8	1.55 7	171.01 6	92	2.88 6	27.09 2	49.70 2
33	36.20 4	1.69 4	177.90 5	93	2.45 0	28.10 9	46.68 1
34	37.40 1	1.83 5	184.26 4	94	2.06 0	29.15 1	43.86 8
435	38.47 7	1.98 2	190.07 7	495	1.72 1	30.22 6	41.29 2
36	39.43 5	2.13 5	195.35 5	96	1.42 1	31.33 5	38.87 1
37	40.28 1	2.29 2	200.11 8	97	1.14 9	32.48 0	36.54 5
38	41.01 4	2.45 5	204.37 7	98	0.91 1	33.66 9	34.35 7
39	41.63 6	2.62 3	208.14 4	99	0.71 0	34.90 8	32.33 3
440	42.14 8	2.79 4	211.43 3	500	0.55 1	36.20 8	30.49 2
41	42.55 3	2.96 7	214.24 9	1	0.42 8	37.57 5	28.78 4
42	42.84 8	3.14 2	216.56 4	2	0.33 3	39.00 4	27.15 1
43	43.02 4	3.31 9	218.34 5	3	0.27 0	40.48 5	25.59 3
44	43.07 6	3.49 9	219.57 1	4	0.24 3	42.00 7	24.11 0
445	43.00 0	3.68 3	220.23 2	505	0.25 9	43.56 8	22.70 9
46	42.82 6	3.87 4	220.47 3	6	0.31 7	45.14 8	21.34 6
47	42.59 4	4.07 0	220.48 7	7	0.41 3	46.72 9	19.98 9
48	42.31 9	4.27 4	220.34 2	8	0.54 9	48.31 0	18.66 3
49	42.01 4	4.48 8	220.10 0	9	0.72 8	49.88 7	17.38 7
450	41.68 9	4.71 2	219.81 3	510	0.95 3	51.45 7	16.18 1
51	41.35 7	4.94 4	219.53 6	11	1.23 2	53.07 1	15.03 9
52	40.99 6	5.18 1	219.15 6	12	1.56 9	54.75 9	13.94 3
53	40.57 4	5.42 3	218.51 0	13	1.95 8	56.50 4	12.90 2
54	40.06 9	5.67 3	217.47 4	14	2.39 6	58.29 1	11.92 8
455	39.46 5	5.93 3	215.95 9	515	2.87 9	60.10 1	11.03 6
56	38.81 3	6.20 1	214.23 4	16	3.41 9	61.91 4	10.22 9
57	38.14 4	6.47 9	212.46 2	17	4.02 3	63.70 8	9.49 0
58	37.43 4	6.76 8	210.49 2	18	4.68 3	65.46 2	8.80 9
59	36.65 9	7.07 0	208.19 2	19	5.39 0	67.16 4	8.17 6
460	35.81 1	7.38 6	205.49 0	520	6.13 1	68.79 9	7.58 3
61	34.97 4	7.71 3	202.84 7	21	6.91 7	70.39 8	7.05 8
62	34.15 4	8.04 6	200.29 0	22	7.76 6	71.99 9	6.61 8
63	33.24 9	8.38 8	197.25 3	23	8.67 2	73.59 9	6.23 6
64	32.18 9	8.74 1	193.32 4	24	9.62 4	75.19 0	5.88 5
465	30.93 3	9.10 9	188.24 8	525	10.61 1	76.76 5	5.53 8
66	29.59 6	9.49 4	182.67 6	26	11.64 0	78.33 2	5.20 5
67	28.27 3	9.90 0	177.16 5	27	12.72 5	79.89 9	4.91 1
68	26.93 9	10.32 9	171.52 2	28	13.85 7	81.45 4	4.64 2
69	25.57 8	10.78 4	165.63 4	29	15.02 5	82.98 4	4.38 7
470	24.18 8	11.26 6	159.45 6	530	16.21 6	84.47 6	4.13 0
71	22.83 1	11.77 1	153.37 0	31	17.42 7	85.93 4	3.87 6
72	21.52 7	12.29 3	147.52 5	32	18.66 8	87.36 7	3.63 8
73	20.23 9	12.83 5	141.66 5	33	19.93 9	88.76 8	3.41 2
74	18.94 7	13.39 5	135.64 1	34	21.23 9	90.12 8	3.19 5
475	17.64 7	13.97 6	129.38 7	535	22.56 7	91.43 9	2.98 3
76	16.38 3	14.57 8	123.22 8	36	23.92 4	92.71 0	2.78 0
77	15.18 5	15.20 3	117.38 4	37	25.31 1	93.95 2	2.59 1
78	14.03 6	15.85 2	111.72 9	38	26.72 7	95.15 5	2.41 2
79	12.92 6	16.52 5	106.18 6	39	28.17 4	96.30 9	2.24 3

TABLE XV (*Continued*)

Wavelength	$E_C\bar{x}$	$E_C\bar{y}$	$E_C\bar{z}$	*Wavelength*	$E_C\bar{x}$	$E_C\bar{y}$	$E_C\bar{z}$
540 $m\mu$	29.64 8	97.40 3	2.07 9	600 $m\mu$	95.25 9	56.60 1	0.06 8
41	31.15 2	98.43 5	1.92 2	1	94.90 5	55.31 9	0.06 5
42	32.68 1	99.40 5	1.77 7	2	94.49 8	54.05 5	0.06 1
43	34.23 3	100.30 8	1.64 1	3	94.02 8	52.80 6	0.05 7
44	35.80 5	101.14 1	1.51 4	4	93.48 3	51.57 0	0.05 4
545	37.39 4	101.89 7	1.39 3	605	92.85 3	50.34 4	0.05 0
46	39.00 0	102.58 3	1.28 0	6	92.14 5	49.13 4	0.04 7
47	40.62 5	103.20 5	1.18 0	7	91.36 9	47.94 3	0.04 3
48	42.26 8	103.76 2	1.08 8	8	90.52 7	46.76 9	0.03 9
49	43.92 7	104.25 2	1.00 3	9	89.61 9	45.60 9	0.03 6
550	45.60 1	104.67 4	0.92 2	610	88.65 1	44.46 5	0.03 3
51	47.28 7	105.02 6	0.84 7	11	87.61 5	43.33 4	0.03 1
52	48.98 4	105.30 6	0.78 0	12	86.50 9	42.21 5	0.02 9
53	50.68 8	105.51 3	0.71 9	13	85.33 3	41.10 6	0.02 7
54	52.39 7	105.64 4	0.66 3	14	84.09 0	40.00 6	0.02 6
555	54.10 7	105.69 6	0.61 0	615	82.78 2	38.91 4	0.02 4
56	55.81 9	105.67 0	0.56 0	16	81.40 5	37.82 9	0.02 3
57	57.52 8	105.56 4	0.51 6	17	79.96 2	36.75 3	0.02 1
58	59.23 2	105.37 9	0.47 6	18	78.45 6	35.68 4	0.02 0
59	60.92 3	105.11 6	0.44 0	19	76.89 1	34.62 1	0.01 9
560	62.59 8	104.77 4	0.40 6	620	75.27 7	33.56 6	0.01 7
61	64.24 9	104.34 9	0.37 5	21	73.59 0	32.50 8	0.01 6
62	65.87 7	103.84 1	0.34 7	22	71.82 1	31.44 5	0.01 5
63	67.48 0	103.25 6	0.32 2	23	69.98 4	30.38 1	0.01 4
64	69.05 9	102.60 2	0.30 0	24	68.09 7	29.32 0	0.01 3
565	70.61 5	101.88 5	0.28 0	625	66.17 4	28.26 7	0.01 2
66	72.14 6	101.10 2	0.26 2	26	64.22 9	27.22 9	0.01 1
67	73.64 9	100.25 4	0.24 6	27	62.27 4	26.21 0	0.01 0
68	75.12 0	99.34 5	0.23 3	28	60.32 8	25.21 6	0.01 0
69	76.56 2	98.38 9	0.22 2	29	58.40 9	24.25 1	0.00 9
570	77.97 1	97.39 0	0.21 1	630	56.53 5	23.32 0	0.00 8
71	79.34 3	96.34 4	0.20 1	31	54.70 1	22.42 1	0.00 7
72	80.67 0	95.24 3	0.19 2	32	52.88 7	21.54 5	0.00 7
73	81.95 2	94.09 4	0.18 5	33	51.10 1	20.69 5	0.00 6
74	83.18 7	92.90 2	0.17 9	34	49.34 0	19.86 8	0.00 5
575	84.37 6	91.67 5	0.17 3	635	47.60 9	19.06 6	0.00 5
76	85.51 7	90.41 2	0.16 7	36	45.90 4	18.28 5	0.00 4
77	86.61 1	89.11 5	0.16 3	37	44.22 0	17.52 5	0.00 4
78	87.65 7	87.79 1	0.15 8	38	42.56 3	16.78 6	0.00 3
79	88.65 7	86.44 6	0.15 5	39	40.92 9	16.06 6	0.00 3
580	89.60 7	85.08 6	0.15 0	640	39.32 2	15.36 5	0.00 3
81	90.50 1	83.70 1	0.14 6	41	37.74 4	14.68 5	0.00 2
82	91.32 1	82.28 0	0.14 2	42	36.20 0	14.02 5	0.00 2
83	92.07 3	80.83 2	0.13 7	43	34.68 8	13.38 6	0.00 2
84	92.76 0	79.36 9	0.13 3	44	33.20 6	12.76 6	0.00 2
585	93.38 1	77.89 5	0.12 9	645	31.75 4	12.16 5	0.00 1
86	93.94 5	76.41 6	0.12 4	46	30.33 5	11.58 2	0.00 1
87	94.45 8	74.93 6	0.12 0	47	28.94 9	11.01 8	0.00 1
88	94.91 6	73.46 0	0.11 6	48	27.59 9	10.47 3	0.00 1
89	95.32 2	71.99 6	0.11 2	49	26.28 5	9.94 7	
590	95.67 8	70.55 2	0.10 8	650	25.00 7	9.43 7	
91	95.97 1	69.11 6	0.10 4	51	23.77 3	8.94 9	
92	96.18 8	67.67 4	0.10 0	52	22.58 4	8.48 1	
93	96.33 0	66.23 6	0.09 6	53	21.44 3	8.03 3	
94	96.39 4	64.80 7	0.09 2	54	20.34 3	7.60 5	
595	96.38 0	63.39 2	0.08 8	655	19.28 3	7.19 4	
96	96.28 1	61.99 3	0.08 4	56	18.26 2	6.79 9	
97	96.10 5	60.60 9	0.08 0	57	17.27 6	6.42 0	
98	95.86 6	59.24 6	0.07 6	58	16.32 1	6.05 5	
99	95.57 8	57.90 8	0.07 2	59	15.39 8	5.70 3	

TABLE XV (*Continued*)

Wavelength	$E_C\bar{x}$	$E_C\bar{y}$	*Wavelength*	$E_C\bar{x}$	$E_C\bar{y}$
660 $m\mu$	14.49 7	5.36 2	700 $m\mu$	0.86 6	0.31 3
61	13.63 1	5.03 5	1	0.80 1	0.28 9
62	12.80 5	4.72 4	2	0.74 2	0.26 8
63	12.01 7	4.42 9	3	0.68 9	0.24 9
64	11.27 3	4.15 0	4	0.64 2	0.23 2
665	10.56 7	3.88 7	705	0.59 9	0.21 6
66	9.89 8	3.63 7	6	0.55 9	0.20 2
67	9.26 3	3.40 0	7	0.52 2	0.18 9
68	8.65 8	3.17 5	8	0.48 7	0.17 6
69	8.08 7	2.96 3	9	0.45 4	0.16 4
670	7.54 3	2.76 2	710	0.42 1	0.15 2
71	7.03 8	2.57 5	11	0.38 9	0.14 1
72	6.58 1	2.40 6	12	0.36 1	0.13 0
73	6.16 7	2.25 3	13	0.33 5	0.12 1
74	5.78 6	2.11 2	14	0.31 1	0.11 2
675	5.43 4	1.98 2	715	0.28 9	0.10 4
76	5.10 6	1.86 1	16	0.26 9	0.09 7
77	4.79 7	1.74 7	17	0.25 0	0.09 0
78	4.50 2	1.63 9	18	0.23 2	0.08 4
79	4.21 3	1.53 3	19	0.21 4	0.07 8
680	3.92 8	1.42 8	720	0.19 8	0.07 2
81	3.65 0	1.32 6	21	0.18 3	0.06 6
82	3.39 0	1.23 1	22	0.16 9	0.06 1
83	3.14 5	1.14 1	23	0.15 6	0.05 7
84	2.91 7	1.05 8	24	0.14 6	0.05 2
685	2.70 3	0.98 0	725	0.13 5	0.04 9
86	2.50 3	0.90 7	26	0.12 5	0.04 5
87	2.31 6	0.83 9	27	0.11 7	0.04 2
88	2.14 0	0.77 4	28	0.10 8	0.03 9
89	1.97 4	0.71 4	29	0.10 0	0.03 6
690	1.81 8	0.65 8	730	0.09 3	0.03 4
91	1.67 5	0.60 6	31	0.08 5	0.03 1
92	1.54 8	0.56 0	32	0.07 9	0.02 9
93	1.43 4	0.51 8	33	0.07 3	0.02 6
94	1.33 2	0.48 1	34	0.06 8	0.02 4
695	1.24 0	0.44 8	735	0.06 3	0.02 3
96	1.15 6	0.41 8	36	0.05 8	0.02 1
97	1.07 8	0.38 9	37	0.05 4	0.01 9
98	1.00 5	0.36 3	38	0.05 0	0.01 8
99	0.93 5	0.33 8	39	0.04 6	0.01 7

CHAPTER VI

DETERMINATION OF TRISTIMULUS VALUES BY THE SELECTED ORDINATE METHOD

In the examples of the preceding chapter, the tristimulus values were computed by a weighted ordinate method. This method involves many multiplications and subsequent addition. Although calculating machines can be used to perform the operations, the procedure under the best of conditions is not only tedious, but, because of the great number of operations, is especially subject to computational errors. The selected ordinate method, described in this chapter, reduces the process of determining the tristimulus values to mere addition. Consequently the operations may be carried out by means of a simple adding machine, preferably one of the type that prints each entry. The printed tape is useful in checking the results for any obvious errors and is also useful as a permanent record.

30. Derivation of Selected Ordinates

The process involved in the determination of the tristimulus values of a colored material is, in the language of the integral calculus, expressed by three integrals of the form

$$X = \int_0^\infty ER\bar{x}d\lambda$$

$$Y = \int_0^\infty ER\bar{y}d\lambda$$

$$Z = \int_0^\infty ER\bar{z}d\lambda$$

where E is a function of wavelength representing the energy distribution of the illuminant, R is a function of wavelength representing the reflection factor (or transmission factor) of the test sample, and $\bar{x}$, $\bar{y}$, and $\bar{z}$ are the tristimulus values for the spectrum colors. These functions are too complex to be formulated analytically. Hence, the indicated integrations must be approximated by the determination of the finite sums:

$$X = \sum_{380}^{780} ER\bar{x}\Delta\lambda$$

$$Y = \sum_{380}^{780} ER\bar{y}\Delta\lambda$$

$$Z = \sum_{380}^{780} ER\bar{z}\Delta\lambda$$

In the weighted ordinate method described in the preceding chapter, a summation is made of values of $ER\bar{x}$, $ER\bar{y}$ and $ER\bar{z}$ respectively at wavelengths which are equally spaced throughout the visible region. In other words, $\Delta\lambda$ has a constant value; and the three products $E\bar{x}$, $E\bar{y}$, and $E\bar{z}$ may be considered to be factors which weight the value of R in the appropriate manner. In the selected ordinate method, the wavelength intervals are unequal, and are so chosen as to give constant values to $E\bar{x}\Delta\lambda$, $E\bar{y}\Delta\lambda$, and $E\bar{z}\Delta\lambda$. The integration process in this case is effected by making a summation of the values of R at the median wavelengths of each of these wavelength intervals. The three sums are multiplied by factors which are proportional to the constants: $E\bar{x}\Delta\lambda$, $E\bar{y}\Delta\lambda$, and $E\bar{z}\Delta\lambda$ respectively. As a matter of convenience, the factors are so chosen as to make the factor for Y equal to the reciprocal of the number of ordinates. The appropriate multiplying factors are given at the foot of each of the succeeding tables. The method of applying the selected ordinate method will be clear from the example below.

The procedure adopted in the determination of wavelengths for the selected ordinate method is as follows: Table XV, for example, gives for Illuminant C the products of E and $\bar{x}$ at wavelength intervals of one millimicron. From this table, a table can be prepared giving the running sum of all the $E\bar{x}$ entries. This is accomplished by adding the entry at 381 millimicrons to the entry at 380 millimicrons. This sum is recorded in the new table opposite the appropriate wavelength designation of 381.5 millimicrons. The entry in Table XV corresponding to 382 millimicrons is then added to the previous sum, and the new total is recorded against the wavelength designation of 382.5 millimicrons. This process is continued for each wavelength up to 780.5. Let the final entry in this table of running sums be represented by S. Then, if 100 selected ordinates are desired, S is multiplied successively by 1/200, 3/200, 5/200 . . . 199/200. These various products are then compared with the table of running sums and, by means of a linear interpolation, the 100 corresponding wavelengths are determined. These are the wavelengths required for the selected ordinate method when 100 ordinates are desired. A similar procedure determines the selected ordinates for $E\bar{y}$ and $E\bar{z}$. The selected ordinates for other illuminants are determined in a corresponding manner.

It is often convenient to have the values of S for the various illuminants. These are listed in the table below.

Values of S

Illuminant	$E\bar{x}$	$E\bar{y}$	$E\bar{z}$
A	11852	10790	3838
B	10369	10465	8922
C	10439	10647	12578

31. Example of the Use of the Selected Ordinate Method

In the selected ordinate method, most of the labor is concentrated in the determination of the appropriate wavelengths, a task which, when once accomplished, need not be repeated. Integration by this method is then exceedingly simple. Consider, for example, the nearly white material selected for illustration on page 34 of the preceding chapter. In this case, 10 selected ordinates are sufficient, and the starred entries in Table XIX are the appropriately selected wavelengths. The values of R corresponding to these wavelengths are the desired selected ordinates. The values are listed in Table XVI. Multiplying the sum of each set of selected ordinates by the proper multiplying factors gives the tristimulus values $X = 0.828$, $Y = 0.847$, and $Z = 0.901$. The trichromatic coefficients are as before

$$x = 0.3209$$

and

$$y = 0.3290.$$

The relative brightness or visual efficiency is in this case given directly by the value of Y as 84.7%.

TABLE XVI

Sample Calculation of Tristimulus Values by the Selected Ordinate Method

Ordinate Number	(X)	(Y)	(Z)
1	0.738	0.804	0.710
2	0.779	0.824	0.725
3	0.845	0.835	0.745
4	0.854	0.843	0.750
5	0.860	0.847	0.760
6	0.865	0.853	0.770
7	0.869	0.857	0.780
8	0.875	0.864	0.785
9	0.876	0.870	0.795
10	0.880	0.876	0.810
	Sum – 8.441	Sum = 8.473	Sum = 7.630
	$X = 0.828$	$Y = 0.847$	$Z = 0.901$

TABLE XVII

Thirty Selected Ordinates for Illuminant A

Ordinate Number	(X)	(Y)	(Z)
1	443.9 61 mμ	487.8 15 mμ	416.4 29 mμ
*2	*516.9 11	*507.7 46	*424.8 74
3	543.9 96	517.2 80	429.4 19
4	554.1 58	524.1 10	432.9 31
*5	561.3 56	*529.7 92	*435.9 70
6	567.1 14	534.8 12	438.7 25
7	572.0 10	539.4 07	441.2 87
*8	*576.3 32	*543.6 97	*443.7 12
9	580.2 46	547.7 66	446.0 41
10	583.8 58	551.6 68	448.2 99
*11	*587.2 46	*555.4 39	*450.4 97
12	590.4 61	559.1 11	452.6 35
13	593.5 42	562.7 11	454.7 24
*14	*596.5 24	*566.2 64	*456.7 82
15	599.4 40	569.7 94	458.8 18
16	602.3 11	573.3 22	460.8 44
*17	*605.1 62	*576.8 70	*462.8 72
18	608.0 17	580.4 61	464.9 22
19	610.9 00	584.1 22	467.0 38
*20	*613.8 35	*587.8 85	*469.2 49
21	616.8 55	591.7 82	471.5 92
22	619.9 97	595.8 52	474.1 03
*23	*623.3 16	*600.1 46	*476.8 45
24	626.8 98	604.7 30	479.9 00
25	630.8 49	609.7 06	483.3 94
*26	*635.3 00	*615.2 19	*487.5 43
27	640.4 83	621.5 24	492.6 62
28	646.9 09	629.2 24	499.2 78
*29	*655.8 71	*639.7 16	*508.3 91
30	673.4 69	658.9 68	526.6 98
Factors (30 ordinates)	*0.03661*	*0.03333*	*0.01185*
Factors (10 ordinates)	*0.10984*	*0.10000*	*0.03555*

TABLE XVIII

Thirty Selected Ordinates for Illuminant B

Ordinate Number	(X)	(Y)	(Z)
1	428.1 26 mμ	472.2 61 mμ	414.7 75 mμ
*2	*442.0 82	*494.4 50	*422.8 99
3	454.1 00	505.7 43	427.1 10
4	468.0 70	513.4 84	430.2 88
*5	*527.7 99	*519.5 66	*432.9 96
6	543.2 77	524.7 63	435.4 45
7	551.9 38	529.4 15	437.7 22
*8	*558.4 96	*533.6 84	*439.8 86
9	563.9 57	537.6 80	441.9 70
10	568.7 51	541.4 74	444.0 00
*11	*573.0 90	*545.1 18	*446.0 00
12	577.1 10	548.6 57	447.9 80
13	580.8 98	552.1 23	449.9 51
*14	*584.5 14	*555.5 45	*451.9 10
15	588.0 05	558.9 51	453.8 63
16	591.4 07	562.3 66	455.8 22
*17	*594.7 49	*565.8 17	*457.8 00
18	598.0 67	569.3 30	459.7 97
19	601.3 88	572.9 34	461.8 30
*20	*604.7 33	*576.6 68	*463.9 07
21	608.1 34	580.5 68	466.0 69
22	611.6 31	584.6 87	468.3 65
*23	*615.2 71	*589.0 93	*470.8 42
24	619.1 22	593.8 61	473.5 54
25	623.2 89	599.1 14	476.6 08
*26	*627.9 56	*605.0 01	*480.1 61
27	633.3 92	611.7 76	484.4 83
28	640.0 65	619.9 37	490.1 82
*29	*649.1 77	*630.9 05	*498.5 87
30	666.2 91	650.7 26	515.2 22
Factors (30 ordinates)	*0.03303*	*0.03333*	*0.02842*
Factors (10 ordinates)	*0.09908*	*0.10000*	*0.08526*

TABLE XIX

Thirty Selected Ordinates for Illuminant C

Ordinate Number	(*X*)	(*Y*)	(*Z*)	*Ordinate Number*	(*X*)	(*Y*)	(*Z*)
1	424.3 98 mμ	465.9 39 mμ	414.1 37 mμ	16	585.0 40 mμ	558.4 80 mμ	453.9 72 mμ
*2	*435.5 33	*489.4 50	*422.1 76	*17	*588.7 28	*561.8 74	*455.9 13
3	443.8 78	500.3 91	426.2 95	18	592.3 60	565.3 23	457.8 89
4	452.1 20	508.6 63	429.3 69	19	595.9 72	568.8 68	459.9 01
*5	*461.2 34	*515.1 50	*431.9 71	*20	*599.6 01	*572.5 39	*461.9 67
6	473.9 74	520.5 97	434.3 20				
7	531.1 73	525.4 19	436.5 03	21	603.2 76	576.3 85	464.0 96
*8	*544.2 79	*529.8 07	*438.5 76	22	607.0 30	580.4 52	466.3 37
9	552.3 95	533.8 80	440.5 71	*23	*610.9 14	*584.8 15	*468.7 47
10	558.7 16	537.7 19	442.5 17	24	614.9 90	589.5 69	471.3 81
				25	619.3 60	594.8 36	474.3 23
*11	*564.0 79	*541.3 82	*444.4 31	*26	*624.1 81	*600.7 98	*477.7 28
12	568.8 58	544.9 19	446.3 35	27	629.7 63	607.7 18	481.8 40
13	573.2 41	548.3 64	448.2 36	28	636.6 09	616.0 95	487.2 02
*14	*577.3 51	*551.7 55	*450.1 41	*29	*645.9 00	*627.2 52	*495.1 58
15	581.2 64	555.1 18	452.0 51	30	663.0 35	647.3 70	511.2 08
Factors (30 Ordinates)	*0.03269*	*0.03333*	*0.03938*	*Factors (10 Ordinates)*	*0.09806*	*0.10000*	*0.11814*

TABLE XX

One Hundred Selected Ordinates for Illuminant A

Ordinate Number	(*X*)	(*Y*)	(*Z*)	*Ordinate Number*	(*X*)	(*Y*)	(*Z*)
1	428.3 38 mμ	467.3 50 mμ	407.6 69 mμ	31	582.6 24 mμ	550.3 19 mμ	447.5 17 mμ
2	442.0 79	485.9 26	415.6 44	32	583.6 84	551.4 76	448.1 88
3	452.9 35	495.1 61	419.4 77	33	584.7 24	552.6 22	448.8 54
4	463.6 43	501.2 86	422.0 31	34	585.7 45	553.7 56	449.5 16
5	478.6 33	505.8 36	424.0 14	35	586.7 50	554.8 80	450.1 70
6	526.7 71	509.4 88	425.6 74	36	587.7 38	555.9 95	450.8 21
7	535.1 32	512.5 74	427.1 16	37	588.7 12	557.1 02	451.4 67
8	540.4 88	515.2 62	428.4 20	38	589.6 71	558.2 00	452.1 05
9	544.6 34	517.6 66	429.6 11	39	590.6 18	559.2 92	452.7 40
10	548.1 14	519.8 56	430.7 22	40	591.5 53	560.3 78	453.3 71
11	551.0 82	521.8 91	431.7 71	41	592.4 76	561.4 57	453.9 97
12	553.7 42	523.8 06	432.7 70	42	593.3 90	562.5 32	454.6 21
13	556.1 46	525.6 14	433.7 27	43	594.2 94	563.6 03	455.2 41
14	558.3 51	527.3 38	434.6 48	44	595.1 91	564.6 70	455.8 59
15	560.3 90	528.9 89	435.5 39	45	596.0 80	565.7 34	456.4 75
16	562.2 91	530.5 81	436.4 02	46	596.9 64	566.7 95	457.0 88
17	564.0 78	532.1 14	437.2 38	47	597.8 42	567.8 55	457.7 01
18	565.7 70	533.6 05	438.0 56	48	598.7 15	568.9 13	458.3 12
19	567.3 78	535.0 50	438.8 56	49	599.5 84	569.9 70	458.9 19
20	568.9 07	536.4 62	439.6 40	50	600.4 49	571.0 28	459.5 27
21	570.3 75	537.8 37	440.4 08	51	601.3 11	572.0 86	460.1 35
22	571.7 81	539.1 84	441.1 60	52	602.1 68	573.1 45	460.7 43
23	573.1 36	540.5 04	441.9 03	53	603.0 25	574.2 05	461.3 51
24	574.4 48	541.7 97	442.6 35	54	603.8 81	575.2 69	461.9 59
25	575.7 13	543.0 68	443.3 55	55	604.7 35	576.3 36	462.5 67
26	576.9 42	544.3 19	444.0 65	56	605.5 90	577.4 04	463.1 77
27	578.1 38	545.5 52	444.7 70	57	606.4 46	578.4 80	463.7 89
28	579.3 00	546.7 66	445.4 67	58	607.3 02	579.5 58	464.4 02
29	580.4 34	547.9 64	446.1 56	59	608.1 60	580.6 42	465.0 26
30	581.5 42	549.1 48	446.8 39	60	609.0 22	581.7 33	465.6 52

TABLE XX (*Continued*)

Ordinate Number	(*X*)	(*Y*)	(*Z*)	*Ordinate Number*	(*X*)	(*Y*)	(*Z*)
61	609.8 87 mμ	582.8 31 mμ	466.2 87 mμ	81	629.4 15 mμ	607.9 13 mμ	482.1 12 mμ
62	610.7 55	583.9 37	466.9 31	82	630.6 39	609.4 45	483.2 08
63	611.6 28	585.0 52	467.5 79	83	631.9 09	611.0 27	484.3 59
64	612.5 05	586.1 77	468.2 40	84	633.2 24	612.6 59	485.5 76
65	613.3 91	587.3 13	468.9 11	85	634.5 90	614.3 50	486.8 71
66	614.2 82	588.4 60	469.5 90	86	636.0 24	616.1 06	488.2 44
67	615.1 81	589.6 20	470.2 85	87	637.5 19	617.9 37	489.7 08
68	616.0 90	590.7 94	470.9 94	88	639.1 01	619.8 52	491.2 71
69	617.0 08	591.9 81	471.7 14	89	640.7 71	621.8 71	492.9 54
70	617.9 38	593.1 85	472.4 48	90	642.5 50	624.0 13	494.7 72
71	618.8 80	594.4 03	473.2 03	91	644.4 64	626.3 06	496.7 44
72	619.8 36	595.6 43	473.9 73	92	646.5 40	628.7 90	498.9 00
73	620.8 07	596.9 04	474.7 62	93	648.8 24	631.4 98	501.2 61
74	621.7 96	598.1 81	475.5 74	94	651.3 65	634.4 92	503.8 65
75	622.8 04	599.4 84	476.4 13	95	654.2 56	637.8 59	506.7 78
76	623.8 35	600.8 13	477.2 80	96	657.6 23	641.7 26	510.1 61
77	624.8 90	602.1 68	478.1 75	97	661.7 23	646.3 50	514.3 34
78	625.9 73	603.5 51	479.1 01	98	667.0 92	652.2 42	519.9 72
79	627.0 86	604.9 70	480.0 64	99	675.1 10	660.6 60	528.3 81
80	628.2 32	606.4 20	481.0 65	100	691.4 94	677.8 02	545.4 45
				Factors	*0.010984*	*0.010000*	*0.003555*

TABLE XXI

One Hundred Selected Ordinates for Illuminant B

Ordinate Number	(*X*)	(*Y*)	(*Z*)	*Ordinate Number*	(*X*)	(*Y*)	(*Z*)
1	418.4 82 mμ	452.0 66 mμ	406.2 79 mμ	31	567.1 34 mμ	540.1 66 mμ	443.2 96 mμ
2	427.1 66	470.3 01	414.0 28	32	568.5 26	541.2 88	443.9 00
3	432.2 61	480.0 57	417.7 09	33	569.8 72	542.3 98	444.5 03
4	436.4 56	486.8 74	420.1 94	34	571.1 84	543.4 94	445.1 03
5	440.2 58	492.1 82	422.0 88	35	572.4 64	544.5 80	445.7 02
6	443.8 78	496.5 19	423.6 52	36	573.7 12	545.6 56	446.2 98
7	447.4 48	500.1 98	424.9 93	37	574.9 33	546.7 17	446.8 92
8	451.0 51	503.3 84	426.1 95	38	576.1 30	547.7 81	447.4 88
9	454.7 18	506.1 88	427.2 89	39	577.3 06	548.8 32	448.0 80
10	458.5 41	508.7 10	428.3 00	40	578.4 61	549.8 77	448.6 72
11	462.6 25	511.0 12	429.2 45	41	579.5 96	550.9 16	449.2 63
12	467.2 14	513.1 41	430.1 41	42	580.7 14	551.9 51	449.8 52
13	473.0 14	515.1 19	430.9 95	43	581.8 16	552.9 81	450.4 42
14	482.3 84	516.9 71	431.8 17	44	582.9 05	554.0 08	451.0 29
15	522.7 53	518.7 23	432.6 12	45	583.9 80	555.0 33	451.6 17
16	531.3 14	520.3 90	433.3 79	46	585.0 44	556.0 56	452.2 04
17	536.5 08	521.9 86	434.1 22	47	586.0 97	557.0 78	452.7 89
18	540.4 85	523.5 26	434.8 49	48	587.1 41	558.0 99	453.3 75
19	543.7 93	525.0 06	435.5 63	49	588.1 76	559.1 21	453.9 61
20	546.6 72	526.4 45	436.2 56	50	589.2 04	560.1 44	454.5 47
21	549.2 36	527.8 36	436.9 39	51	590.2 24	561.1 68	455.1 35
22	551.5 71	529.1 91	437.6 12	52	591.2 37	562.1 94	455.7 24
23	553.7 18	530.5 13	438.2 72	53	592.2 46	563.2 24	456.3 14
24	555.7 18	531.8 01	438.9 23	54	593.2 50	564.2 59	456.9 07
25	557.5 96	533.0 62	439.5 68	55	594.2 50	565.2 95	457.5 02
26	559.3 71	534.2 98	440.2 03	56	595.2 47	566.3 41	458.0 98
27	561.0 55	535.5 12	440.8 32	57	596.2 44	567.3 88	458.6 95
28	562.6 67	536.7 03	441.4 56	58	597.2 38	568.4 44	459.2 94
29	564.2 11	537.8 74	442.0 72	59	598.2 33	569.5 07	459.8 98
30	565.6 99	539.0 28	442.6 86	60	599.2 29	570.5 79	460.5 03

TABLE XXI (*Continued*)

Ordinate Number	(*X*)	(*Y*)	(*Z*)	*Ordinate Number*	(*X*)	(*Y*)	(*Z*)
61	600.2 24 mμ	571.6 60 mμ	461.1 14 mμ	81	621.7 86 mμ	597.2 12 mμ	475.4 91 mμ
62	601.2 21	572.7 52	461.7 27	82	623.0 72	598.8 38	476.4 44
63	602.2 20	573.8 55	462.3 43	83	624.4 00	600.5 16	477.4 40
64	603.2 22	574.9 70	462.9 65	84	625.7 79	602.2 60	478.4 84
65	604.2 28	576.0 98	463.5 90	85	627.2 14	604.0 68	479.5 85
66	605.2 38	577.2 41	464.2 24	86	628.7 14	605.9 52	480.7 50
67	606.2 54	578.4 00	464.8 66	87	630.2 86	607.9 19	481.9 88
68	607.2 77	579.5 75	465.5 15	88	631.9 44	609.9 78	483.3 08
69	608.3 06	580.7 69	466.1 82	89	633.6 91	612.1 45	484.7 32
70	609.3 45	581.9 83	466.8 57	90	635.5 49	614.4 34	486.2 79
71	610.3 94	583.2 17	467.5 43	91	637.5 38	616.8 71	487.9 77
72	611.4 52	584.4 75	468.2 47	92	639.6 88	619.4 80	489.8 50
73	612.5 25	585.7 59	468.9 67	93	642.0 32	622.3 14	491.9 13
74	613.6 11	587.0 70	469.7 02	94	644.6 23	625.4 35	494.3 07
75	614.7 14	588.4 10	470.4 54	95	647.5 48	628.9 55	497.0 48
76	615.8 34	589.7 83	471.2 32	96	650.9 43	633.0 11	500.2 75
77	616.9 74	591.1 88	472.0 31	97	655.0 22	637.8 36	504.2 02
78	618.1 36	592.6 30	472.8 52	98	660.2 35	643.9 04	509.2 06
79	619.3 22	594.1 13	473.6 99	99	667.8 42	652.4 27	516.8 19
80	620.5 37	595.6 38	474.5 76	100	683.3 76	668.9 78	533.3 14
				Factors	*0.009908*	*0.010000*	*0.008526*

TABLE XXII

ONE HUNDRED SELECTED ORDINATES FOR ILLUMINANT C

Ordinate Number	(*X*)	(*Y*)	(*Z*)	*Ordinate Number*	(*X*)	(*Y*)	(*Z*)
1	415.3 55 mμ	446.6 43 mμ	405.6 78 mμ	31	556.6 41 mμ	536.4 00 mμ	441.8 39 mμ
2	423.5 53	463.9 88	413.4 15	32	558.4 27	537.5 33	442.4 20
3	427.8 98	473.8 21	417.0 33	33	560.1 27	538.6 50	442.9 97
4	431.2 53	480.7 18	419.5 13	34	561.7 54	539.7 54	443.5 72
5	434.1 68	486.1 30	421.3 82	35	563.3 19	540.8 42	444.1 45
6	436.8 50	490.6 38	422.9 08	36	564.8 28	541.9 20	444.7 17
7	439.3 94	494.4 88	424.2 30	37	566.2 90	542.9 86	445.2 89
8	441.8 57	497.8 60	425.4 07	38	567.7 07	544.0 46	445.8 59
9	444.2 80	500.8 67	426.4 71	39	569.0 85	545.0 92	446.4 30
10	446.7 16	503.5 74	427.4 51	40	570.4 32	546.1 34	447.0 00
11	449.1 85	506.0 33	428.3 65	41	571.7 44	547.1 67	447.5 70
12	451.6 95	508.3 01	429.2 28	42	573.0 30	548.1 93	448.1 42
13	454.2 70	510.4 14	430.0 51	43	574.2 90	549.2 17	448.7 13
14	456.9 45	512.3 92	430.8 40	44	575.5 30	550.2 36	449.2 85
15	459.7 57	514.2 57	431.6 04	45	576.7 48	551.2 50	449.8 56
16	462.7 62	516.0 20	432.3 41	46	577.9 48	552.2 61	450.4 28
17	466.0 64	517.6 96	433.0 54	47	579.1 32	553.2 70	451.0 01
18	469.9 44	519.3 04	433.7 51	48	580.3 02	554.2 80	451.5 74
19	474.9 20	520.8 48	434.4 33	49	581.4 57	555.2 87	452.1 49
20	482.8 48	522.3 40	435.0 97	50	582.6 00	556.2 95	452.7 23
21	520.9 35	523.7 85	435.7 52	51	583.7 32	557.3 02	453.2 99
22	530.1 51	525.1 87	436.3 96	52	584.8 54	558.3 12	453.8 76
23	535.4 51	526.5 51	437.0 27	53	585.9 70	559.3 25	454.4 54
24	539.4 64	527.8 76	437.6 53	54	587.0 78	560.3 40	455.0 36
25	542.7 90	529.1 72	438.2 67	55	588.1 80	561.3 61	455.6 19
26	545.6 74	530.4 35	438.8 76	56	589.2 75	562.3 85	456.2 07
27	548.2 45	531.6 74	439.4 87	57	590.3 70	563.4 15	456.7 97
28	550.5 89	532.8 87	440.0 77	58	591.4 57	564.4 54	457.3 92
29	552.7 43	534.0 77	440.6 69	59	592.5 42	565.4 97	457.9 89
30	554.7 52	535.2 47	441.2 56	60	593.6 24	566.5 50	458.5 87

TABLE XXII (*Continued*)

Ordinate Number	(*X*)	(*Y*)	(*Z*)	*Ordinate Number*	(*X*)	(*Y*)	(*Z*)
61	594.7 07 $m\mu$	567.6 13 $m\mu$	459.1 93 $m\mu$	81	617.7 89 $m\mu$	592.9 26 $m\mu$	473.2 52 $m\mu$
62	595.7 92	568.6 87	459.7 99	82	619.1 32	594.5 55	474.1 69
63	596.8 77	569.7 72	460.4 12	83	620.5 12	596.2 52	475.1 23
64	597.9 64	570.8 69	461.0 30	84	621.9 38	598.0 16	476.1 26
65	599.0 55	571.9 80	461.6 53	85	623.4 16	599.8 50	477.1 80
66	600.1 50	573.1 05	462.2 81	86	624.9 62	601.7 67	478.2 90
67	601.2 50	574.2 45	462.9 14	87	626.5 73	603.7 72	479.4 63
68	602.3 51	575.4 05	463.5 53	88	628.2 72	605.8 79	480.7 21
69	603.4 60	576.5 80	464.2 05	89	630.0 71	608.0 98	482.0 71
70	604.5 78	577.7 80	464.8 64	90	631.9 82	610.4 46	483.5 24
71	605.7 04	579.0 00	465.5 33	91	634.0 24	612.9 50	485.1 23
72	606.8 40	580.2 44	466.2 22	92	636.2 27	615.6 29	486.8 88
73	607.9 88	581.5 11	466.9 24	93	638.6 22	618.5 32	488.8 60
74	609.1 48	582.8 10	467.6 37	94	641.2 72	621.7 18	491.0 96
75	610.3 21	584.1 40	468.3 71	95	644.2 48	625.2 82	493.6 85
76	611.5 10	585.4 99	469.1 26	96	647.6 86	629.3 95	496.7 71
77	612.7 19	586.9 00	469.9 00	97	651.8 01	634.3 07	500.5 69
78	613.9 50	588.3 38	470.6 96	98	657.0 37	640.4 69	505.4 71
79	615.2 02	589.8 20	471.5 17	99	664.5 47	649.0 90	512.7 02
80	616.4 79	591.3 45	472.3 70	100	679.8 08	665.5 45	528.9 77
				Factors	*0.009806*	*0.010000*	*0.011814*

CHAPTER VII

TRICHROMATIC COEFFICIENTS

TABLE XII in Chapter V provides all the information necessary for the computation of the trichromatic coefficients of the spectrum colors. The appropriate formula are:

$$x = \frac{\bar{x}}{\bar{x} + \bar{y} + \bar{z}} \qquad (17a)$$

$$y = \frac{\bar{y}}{\bar{x} + \bar{y} + \bar{z}} \qquad (17b)$$

$$z = \frac{\bar{z}}{\bar{x} + \bar{y} + \bar{z}} \qquad (17c)$$

Ordinarily only x and y are required because $z = 1 - (x + y)$. Despite the fact that x, y, and z can be determined by computation from the tables of $\bar{x}$, $\bar{y}$, and $\bar{z}$, it is nevertheless a convenience to have the trichromatic coefficients readily available. Values of x, y, and z are listed in Table XXIV for wavelength intervals of one millimicron. The method by which this table was interpolated from the I.C.I. data was outlined in Chapter V. As in previous tables, the figures following the space should be ignored except when Table XXIV is used as a basis for the construction of other tables.

Table XXIII gives the trichromatic coefficients of a few illuminants of special interest.

It will be noticed that the trichromatic coefficients of Illuminant A are identical with those of a black body at 2848° K. Illuminant B and Illuminant C correspond approximately to temperatures of 4800° K and 6500° K respectively.

TABLE XXIII

Trichromatic Coefficients of Important Illuminants

Source	x	y
Black Body at:		
1000° K	0.6524	0.3448
1500°	0.5852	0.3934
1900°	0.5372	0.4114
2360°	0.4893	0.4150
2848°	0.4476	0.4075
3500°	0.4049	0.3906
4800°	0.3506	0.3560
6500°	0.3133	0.3235
10,000°	0.2796	0.2871
24,000°	0.2532	0.2532
∞	0.2399	0.2342
I.C.I. Illuminant A	0.4476	0.4075
I.C.I. Illuminant B	0.3485	0.3518
I.C.I. Illuminant C	0.3101	0.3163
Mean Noon Sunlight	0.3442	0.3534
Sun outside atmosphere	0.3204	0.3301

TABLE XXIV

Trichromatic Coefficients of the Spectrum Colors

Wavelength	x	y	z
380 mμ	0.1741 0000	0.0050 0000	0.8209 0000
81	0.1740 8080	0.0049 8936	0.8209 2984
82	0.1740 5920	0.0049 7928	0.8209 6152
83	0.1740 3520	0.0049 6952	0.8209 9528
84	0.1740 0880	0.0049 5984	0.8210 3136
385	0.1739 8000	0.0049 5000	0.8210 7000
86	0.1739 4848	0.0049 4016	0.8211 1136
87	0.1739 1424	0.0049 3048	0.8211 5528
88	0.1738 7776	0.0049 2072	0.8212 0152
89	0.1738 3952	0.0049 1064	0.8212 4984
390	0.1738 0000	0.0049 0000	0.8213 0000
91	0.1737 5952	0.0048 8832	0.8213 5216
92	0.1737 1770	0.0048 7582	0.8214 0648
93	0.1736 7420	0.0048 6308	0.8214 6272
94	0.1736 2840	0.0048 5096	0.8215 2064
395	0.1735 8000	0.0048 4000	0.8215 8000
96	0.1735 2832	0.0048 3040	0.8216 4128
97	0.1734 7376	0.0048 2160	0.8217 0464
98	0.1734 1700	0.0048 1364	0.8217 6936
99	0.1733 5888	0.0048 0640	0.8218 3472
400	0.1733 0000	0.0048 0000	0.8219 0000
1	0.1732 4120	0.0047 9408	0.8219 6472
2	0.1731 8200	0.0047 8864	0.8220 2936
3	0.1731 2120	0.0047 8416	0.8220 9464
4	0.1730 5760	0.0047 8112	0.8221 6128
405 mμ	0.1729 9000	0.0047 8000	0.8222 3000
6	0.1729 1888	0.0047 8176	0.8222 9936
7	0.1728 4504	0.0047 8608	0.8223 6888
8	0.1727 6776	0.0047 9152	0.8224 4072
9	0.1726 8632	0.0047 9664	0.8225 1704
410	0.1726 0000	0.0048 0000	0.8226 0000
11	0.1725 1024	0.0047 9648	0.8226 9328
12	0.1724 1752	0.0047 8704	0.8227 9544
13	0.1723 1968	0.0047 7936	0.8229 0096
14	0.1722 1456	0.0047 8112	0.8230 0432
415	0.1721 0000	0.0048 0000	0.8231 0000
16	0.1719 7920	0.0048 3440	0.8231 8640
17	0.1718 5360	0.0048 7920	0.8232 6720
18	0.1717 1840	0.0049 3680	0.8233 4480
19	0.1715 6880	0.0050 0960	0.8234 2160
420	0.1714 0000	0.0051 0000	0.8235 0000
21	0.1712 104	0.0052 0800	0.8235 8160
22	0.1710 032	0.0053 3200	0.8236 6480
23	0.1707 808	0.0054 7200	0.8237 4720
24	0.1705 456	0.0056 2800	0.8238 2640
425	0.1703 000	0.0058 0000	0.8239 0000
26	0.1700 488	0.0059 8480	0.8239 6640
27	0.1697 904	0.0061 8240	0.8240 2720
28	0.1695 176	0.0063 9760	0.8240 8480
29	0.1692 232	0.0066 3520	0.8241 4160

TABLE XXIV *(Continued)*

Wavelength	x	y	z	*Wavelength*	x	y	z
430 mμ	0.1689 000	0.0069 0000	0.8242 0000	490 mμ	0.0454 000	0.2950 000	0.6596 000
31	0.1685 464	0.0071 9200	0.8242 6160	91	0.0408 248	0.3169 144	0.6422 608
32	0.1681 672	0.0075 0800	0.8243 2480	92	0.0362 224	0.3400 112	0.6237 664
33	0.1677 648	0.0078 4800	0.8243 8720	93	0.0317 176	0.3639 208	0.6043 616
34	0.1673 416	0.0082 1200	0.8244 4640	94	0.0274 352	0.3882 736	0.5842 912
435	0.1669 000	0.0086 0000	0.8245 0000	495	0.0235 000	0.4127 000	0.5638 000
36	0.1664 448	0.0090 1200	0.8245 4320	96	0.0198 416	0.4374 768	0.5426 816
37	0.1659 744	0.0094 4800	0.8245 7760	97	0.0163 768	0.4628 504	0.5207 728
38	0.1654 816	0.0099 0800	0.8246 1040	98	0.0132 112	0.4884 056	0.4983 832
39	0.1649 592	0.0103 9200	0.8246 4880	99	0.0104 504	0.5137 272	0.4758 224
440	0.1644 000	0.0109 0000	0.8247 0000	500	0.0082 000	0.5384 000	0.4534 000
41	0.1638 104	0.0114 2560	0.8247 6400	1	0.0064 072	0.5626 112	0.4309 816
42	0.1631 952	0.0119 6880	0.8248 3600	2	0.0050 016	0.5866 376	0.4083 608
43	0.1625 448	0.0125 3920	0.8249 1600	3	0.0040 624	0.6101 984	0.3857 392
44	0.1618 496	0.0131 4640	0.8250 0400	4	0.0036 688	0.6330 128	0.3633 184
445	0.1611 000	0.0138 0000	0.8251 0000	505	0.0039 000	0.6548 000	0.3413 000
46	0.1602 944	0.0144 9840	0.8252 0720	6	0.0047 448	0.6757 616	0.3194 936
47	0.1594 392	0.0152 3520	0.8253 2560	7	0.0061 504	0.6960 848	0.2977 648
48	0.1585 368	0.0160 1280	0.8254 5040	8	0.0081 336	0.7154 672	0.2763 992
49	0.1575 896	0.0168 3360	0.8255 7680	9	0.0107 112	0.7336 064	0.2556 824
450	0.1566 000	0.0177 0000	0.8257 0000	510	0.0139 000	0.7502 000	0.2359 000
51	0.1555 728	0.0185 9760	0.8258 2960	11	0.0177 736	0.7653 504	0.2168 760
52	0.1545 064	0.0195 2480	0.8259 6880	12	0.0223 208	0.7792 592	0.1984 200
53	0.1533 936	0.0205 0320	0.8261 0320	13	0.0274 312	0.7917 728	0.1807 960
54	0.1522 272	0.0215 5440	0.8262 1840	14	0.0329 944	0.8027 376	0.1642 680
455	0.1510 000	0.0227 0000	0.8263 0000	515	0.0389 000	0.8120 000	0.1491 000
56	0.1497 136	0.0239 2080	0.8263 6560	16	0.0452 424	0.8193 904	0.1353 672
57	0.1483 728	0.0252 0240	0.8264 2480	17	0.0520 952	0.8250 112	0.1228 936
58	0.1469 752	0.0265 7360	0.8264 5120	18	0.0593 168	0.8291 168	0.1115 664
59	0.1455 184	0.0280 6320	0.8264 1840	19	0.0667 656	0.8319 616	0.1012 728
460	0.1440 000	0.0297 0000	0.8263 0000	520	0.0743 000	0.8338 000	0.0919 000
61	0.1424 424	0.0314 120	0.8261 456	21	0.0819 824	0.8343 648	0.0836 528
62	0.1408 472	0.0331 800	0.8259 728	22	0.0899 072	0.8334 864	0.0766 064
63	0.1391 808	0.0351 120	0.8257 072	23	0.0979 808	0.8315 656	0.0704 536
64	0.1374 096	0.0373 160	0.8252 744	24	0.1061 096	0.8290 032	0.0648 872
465	0.1355 000	0.0399 000	0.8246 000	525	0.1142 000	0.8262 000	0.0596 000
66	0.1334 552	0.0428 096	0.8237 352	26	0.1222 984	0.8230 168	0.0546 848
67	0.1312 976	0.0459 728	0.8227 296	27	0.1304 672	0.8191 864	0.0503 464
68	0.1290 224	0.0494 712	0.8215 064	28	0.1386 368	0.8149 176	0.0464 456
69	0.1266 248	0.0533 864	0.8199 888	29	0.1467 376	0.8104 192	0.0428 432
470	0.1241 000	0.0578 000	0.8181 000	530	0.1547 000	0.8059 000	0.0394 000
71	0.1214 592	0.0626 192	0.8159 216	31	0.1625 112	0.8013 440	0.0361 448
72	0.1187 056	0.0677 896	0.8135 048	32	0.1702 176	0.7966 120	0.0331 704
73	0.1158 224	0.0734 504	0.8107 272	33	0.1778 384	0.7917 280	0.0304 336
74	0.1127 928	0.0797 408	0.8074 664	34	0.1853 928	0.7867 160	0.0278 912
475	0.1096 000	0.0868 000	0.8036 000	535	0.1929 000	0.7816 000	0.0255 000
76	0.1062 520	0.0945 448	0.7992 032	36	0.2003 440	0.7763 752	0.0232 808
77	0.1027 600	0.1028 824	0.7943 576	37	0.2077 120	0.7710 256	0.0212 624
78	0.0991 120	0.1119 376	0.7889 504	38	0.2150 280	0.7655 584	0.0194 136
79	0.0952 960	0.1218 352	0.7828 688	39	0.2223 160	0.7599 808	0.0177 032
480	0.0913 000	0.1327 000	0.7760 000	540	0.2296 000	0.7543 000	0.0161 000
81	0.0870 664	0.1444 648	0.7684 688	41	0.2368 784	0.7485 0480	0.0146 1680
82	0.0826 032	0.1570 464	0.7603 504	42	0.2441 352	0.7425 9040	0.0132 7440
83	0.0779 968	0.1705 456	0.7514 576	43	0.2513 728	0.7365 7360	0.0120 5360
84	0.0733 336	0.1850 632	0.7416 032	44	0.2585 936	0.7304 7120	0.0109 3520
485	0.0687 000	0.2007 000	0.7306 000	545	0.2658 000	0.7243 0000	0.0099 0000
86	0.0640 624	0.2175 024	0.7184 352	46	0.2729 872	0.7180 5040	0.0089 6240
87	0.0593 632	0.2354 032	0.7052 336	47	0.2801 536	0.7117 1120	0.0081 3520
88	0.0546 528	0.2543 328	0.6910 144	48	0.2873 064	0.7052 9680	0.0073 9680
89	0.0499 816	0.2742 216	0.6757 968	49	0.2944 528	0.6988 2160	0.0067 2560

TABLE XXIV (*Continued*)

Wavelength	*x*	*y*	*z*
550 mμ	0.3016 000	0.6923 0000	0.0061 0000
51	0.3087 448	0.6857 2592	0.0055 2928
52	0.3158 824	0.6790 8976	0.0050 2784
53	0.3230 176	0.6724 0064	0.0045 8176
54	0.3301 552	0.6656 6768	0.0041 7712
555	0.3373 000	0.6589 0000	0.0038 0000
56	0.3444 568	0.6520 8736	0.0034 5584
57	0.3516 224	0.6452 2368	0.0031 5392
58	0.3587 896	0.6383 2432	0.0028 8608
59	0.3659 512	0.6314 0464	0.0026 4416
560	0.3731 000	0.6244 8000	0.0024 2000
61	0.3802 360	0.6175 4720	0.0022 1680
62	0.3873 640	0.6105 9600	0.0020 4000
63	0.3944 840	0.6036 3120	0.0018 8480
64	0.4015 960	0.5966 5760	0.0017 4640
565	0.4087 000	0.5896 8000	0.0016 2000
66	0.4158 040	0.5826 8752	0.0015 0848
67	0.4229 080	0.5756 7696	0.0014 1504
68	0.4300 000	0.5686 6464	0.0013 3536
69	0.4370 680	0.5616 6688	0.0012 6512
570	0.4441 000	0.5547 0000	0.0012 0000
71	0.4511 008	0.5477 5760	0.0011 4160
72	0.4580 784	0.5408 2880	0.0010 9280
73	0.4650 256	0.5339 2320	0.0010 5120
74	0.4719 352	0.5270 5040	0.0010 1440
575	0.4788 000	0.5202 2000	0.0009 8000
76	0.4856 264	0.5134 2416	0.0009 4944
77	0.4924 192	0.5066 5648	0.0009 2432
78	0.4991 688	0.4999 2872	0.0009 0248
79	0.5058 656	0.4932 5264	0.0008 8176
580	0.5125 000	0.4866 4000	0.0008 6000
81	0.5190 800	0.4800 8280	0.0008 3720
82	0.5256 120	0.4735 7320	0.0008 1480
83	0.5320 840	0.4671 2320	0.0007 9280
84	0.5384 840	0.4607 4480	0.0007 7120
585	0.5448 000	0.4544 5000	0.0007 5000
86	0.5510 448	0.4482 2584	0.0007 2936
87	0.5572 264	0.4420 6432	0.0007 0928
88	0.5633 256	0.4359 8488	0.0006 8952
89	0.5693 232	0.4300 0696	0.0006 6984
590	0.5752 000	0.4241 5000	0.0006 5000
91	0.5809 704	0.4183 9960	0.0006 3000
92	0.5866 472	0.4127 4280	0.0006 1000
93	0.5922 088	0.4072 0120	0.0005 9000
94	0.5976 336	0.4017 9640	0.0005 7000
595	0.6029 000	0.3965 5000	0.0005 5000
96	0.6079 968	0.3914 7320	0.0005 3000
97	0.6129 384	0.3865 5160	0.0005 1000
98	0.6177 416	0.3817 6840	0.0004 9000
99	0.6224 232	0.3771 0680	0.0004 7000
600	0.6270 000	0.3725 5000	0.0004 5000
1	0.6314 832	0.3680 8680	0.0004 3000
2	0.6358 616	0.3637 2840	0.0004 1000
3	0.6401 184	0.3594 9160	0.0003 9000
4	0.6442 368	0.3553 9320	0.0003 7000
605	0.6482 000	0.3514 5000	0.0003 5000
6	0.6520 032	0.3476 6760	0.0003 2920
7	0.6556 576	0.3440 3480	0.0003 0760
8	0.6591 704	0.3405 4320	0.0002 8640
9	0.6625 488	0.3371 8440	0.0002 6680
610 mμ	0.6658 000	0.3339 5000	0.0002 5000
11	0.6689 176	0.3308 4576	0.0002 3664
12	0.6718 968	0.3278 7728	0.0002 2592
13	0.6747 472	0.3250 3592	0.0002 1688
14	0.6774 784	0.3223 1304	0.0002 0856
615	0.6801 000	0.3197 0000	0.0002 0000
16	0.6826 024	0.3172 0640	0.0001 9120
17	0.6849 792	0.3148 3800	0.0001 8280
18	0.6872 448	0.3125 8040	0.0001 7480
19	0.6894 136	0.3104 1920	0.0001 6720
620	0.6915 000	0.3083 4000	0.0001 6000
21	0.6934 960	0.3063 5064	0.0001 5336
22	0.6953 920	0.3044 6072	0.0001 4728
23	0.6972 000	0.3026 5848	0.0001 4152
24	0.6989 320	0.3009 3216	0.0001 3584
625	0.7006 000	0.2992 7000	0.0001 3000
26	0.7021 944	0.2976 8160	0.0001 2400
27	0.7037 072	0.2961 7480	0.0001 1800
28	0.7051 528	0.2947 3520	0.0001 1200
29	0.7065 456	0.2933 4840	0.0001 0600
630	0.7079 000	0.2920 0000	0.0001 0000
31	0.7092 144	0.2906 9176	0.0000 9384
32	0.7104 792	0.2894 3328	0.0000 8752
33	0.7116 968	0.2882 2192	0.0000 8128
34	0.7128 696	0.2870 5504	0.0000 7536
635	0.7140 000	0.2859 3000	0.0000 7000
36	0.7150 864	0.2848 4824	0.0000 6536
37	0.7161 272	0.2838 1152	0.0000 6128
38	0.7171 248	0.2828 1768	0.0000 5752
39	0.7180 816	0.2818 6456	0.0000 5384
640	0.7190 000	0.2809 5000	0.0000 5000
41	0.7198 800	0.2800 7400	0.0000 4600
42	0.7207 200	0.2792 3800	0.0000 4200
43	0.7215 200	0.2784 4200	0.0000 3800
44	0.7222 800	0.2776 8600	0.0000 3400
645	0.7230 000	0.2769 7000	0.0000 3000
46	0.7236 752	0.2762 9896	0.0000 2584
47	0.7243 056	0.2756 7288	0.0000 2152
48	0.7248 984	0.2750 8432	0.0000 1728
49	0.7254 608	0.2745 2584	0.0000 1336
650	0.7260 000	0.2739 9000	0.0000 1000
51	0.7265 144	0.2734 7856	0.0000 0704
52	0.7269 992	0.2729 9648	0.0000 0432
53	0.7274 568	0.2725 4112	0.0000 0208
54	0.7278 896	0.2721 0984	0.0000 0056
655	0.7283 000	0.2717 0000	
56	0.7286 880	0.2713 1200	
57	0.7290 520	0.2709 4800	
58	0.7293 920	0.2706 0800	
59	0.7297 080	0.2702 9200	
660	0.7300 000	0.2700 0000	
61	0.7302 6160	0.2697 3840	
62	0.7304 9280	0.2695 0720	
63	0.7307 0320	0.2692 9680	
64	0.7309 0240	0.2690 9760	
665	0.7311 0000	0.2689 0000	
66	0.7312 9600	0.2687 0400	
67	0.7314 8400	0.2685 1600	
68	0.7316 6400	0.2683 3600	
69	0.7318 3600	0.2681 6400	

TABLE XXIV (*Continued*)

Wavelength	x	y	*Wavelength*	x	y
670 mμ	0.7320 0000	0.2680 0000	686 mμ	0.7340 6848	0.2659 3152
71	0.7321 5280	0.2678 4720	87	0.7341 6264	0.2658 3736
72	0.7322 9440	0.2677 0560	88	0.7342 5056	0.2657 4944
73	0.7324 2960	0.2675 7040	89	0.7343 3032	0.2656 6968
74	0.7325 6320	0.2674 3680	690	0.7344 0000	0.2656 0000
675	0.7327 0000	0.2673 0000	91	0.7344 5848	0.2655 4152
76	0.7328 4208	0.2671 5792	92	0.7345 0704	0.2654 9296
77	0.7329 8624	0.2670 1376	93	0.7345 4736	0.2654 5264
78	0.7331 2936	0.2668 7064	94	0.7345 8112	0.2654 1888
79	0.7332 6832	0.2667 3168	695	0.7346 1000	0.2653 9000
680	0.7334 0000	0.2666 0000	96	0.7346 3352	0.2653 6648
81	0.7335 2456	0.2664 7544	97	0.7346 5056	0.2653 4944
82	0.7336 4408	0.2663 5592	98	0.7346 6184	0.2653 3816
83	0.7337 5832	0.2662 4168	99	0.7346 6808	0.2653 3192
84	0.7338 6704	0.2661 3296	700	0.7346 7000	0.2653 3000
685	0.7339 7000	0.2660 3000	780	0.7346 7000	0.2653 3000

CHAPTER VIII

GRAPHICAL REPRESENTATION OF COLORIMETRIC DATA

IT WAS shown in Chapter I that the only fundamental color language is one that is based on either spectroradiometric or spectrophotometric data, the former being applicable chiefly to illuminants and the latter to transparent or opaque materials. Because the eye is not an analytical instrument, a spectroradiometric or spectrophotometric specification furnishes more information than the unaided eye is capable of acquiring. Hence, the subsequent chapters of this volume have been concerned largely with methods for evaluating a stimulus that the eye accepts as equivalent under certain standardized conditions. Although the tristimulus values provide a specification of an equivalent stimulus, the results of a color measurement are more readily interpreted if the tristimulus values are converted into brightness (visual efficiency), dominant wavelength, and purity. To facilitate this conversion, a set of charts based on the I.C.I. observer and Illuminant C has been prepared. These charts will be found on pages 62 to 84. The method of employing these charts is simply to determine the trichromatic coefficients, x and y, from the tristimulus values, X, Y, and Z. The approximate location of the point determined by these trichromatic coefficients is ascertained by reference to the key chart on page 61. The exact location is then spotted on the large scale chart corresponding to the area indicated on the key chart. This large scale chart contains lines of constant dominant wavelength radiating from a point whose coördinates are the trichromatic coefficients of Illuminant C. It contains also contour lines of excitation purity. In this way, the dominant wavelength and excitation purity may be interpolated directly from the chart.

The use of these charts will be made clear by considering again two of the examples discussed in previous chapters. The example on page 10 concerned a surface of green paint whose trichromatic coefficients were $x = 0.2487$ and $y = 0.3881$. Reference to chart No. 8 shows that the dominant wavelength of this sample is 506.0 millimicrons and the excitation purity is 20%. The relative brightness, or visual efficiency, was previously shown to be 24.2%. These three parameters completely evaluate the color of the test sample under the assumed conditions (I.C.I. observer and Illuminant C).

The example on page 50 was to determine the trichromatic coefficients of a sheet of nearly white paper. The trichromatic coefficients were found to be $x = 0.3209$ and $y = 0.3290$. Because nearly white materials are so very common, a portion of chart No. 12 is shown on an expanded scale in chart No. 12a. From this chart, it will be seen that the point to which these values of x and y correspond has a dominant wavelength of 575 millimicrons and an excitation purity of 6.2%. The relative brightness, or visual efficiency, was previously shown to be 84.7%. As before, these three parameters completely evaluate the color of the test sample under the assumed conditions.

It must be remembered that these charts are based on the assumption that the illuminant has trichromatic coefficients identical with those of Illuminant C. Although problems occasionally arise where the illumination can not be assumed to have the chromaticity of this illuminant,[1] the vast majority of problems are adequately solved with Illuminant C as a basis. There is a distinct advantage in having all results, whenever possible, expressed in terms of the same basic illuminant. Furthermore, to include charts based on all the illuminants that might possibly be encountered, would obviously be prohibitive. It should also be emphasized that excitation purity has been used throughout this volume in preference to colorimetric purity. Excitation purity was adopted because its values are always positive, and the purity contours on a chromaticity diagram are continuous. If colorimetric purity is used, there is a discontinuity in the purity contours and a crowding of the contour lines in certain wavelength regions that is highly objectionable. Since some work has been done in the past on the basis of colorimetric purity, it is desirable to have formulae for interconversion. Excitation purity is defined by the equation

$$p_e = \frac{y_s - y_w}{y_l - y_w} = \frac{x_s - x_w}{x_l - x_w} \qquad (18)$$

where x_s, y_s are the trichromatic coefficients of the sample, x_w, y_w are the trichromatic coefficients of the illuminant, x_l, y_l are the coefficients of the point on the spectrum locus where the line from the white point extended through the sample point intersects the spectrum locus or the line joining its extremities. Consequently, excitation purity is unity for every point on the boundary of realizable colors. Colorimetric purity is defined by the equation

$$p_c = \frac{y_l}{y_s} \cdot \frac{y_s - y_w}{y_l - y_w} = \frac{y_l}{y_s} \cdot \frac{x_s - x_w}{x_l - x_w} \qquad (19)$$

where the symbols have the same meaning as above, except that x_l, y_l are always the trichromatic coefficients of a point on the spectrum locus. In the case of purples, this point is the one representing the spectral color complementary to the sample. Consequently colorimetric purity can be unity only for points on the spectrum locus.

[1] A typical problem of this sort is a specification of the color of railway signal glasses.

Except in the case of colors that are purple, elimination of y_s between equations (18) and (19) gives the transformation equations

$$p_c = \frac{y_l p_e}{y_w + p_e (y_l - y_w)} \qquad (20a)$$

$$p_e = \frac{y_w p_c}{y_l - p_c (y_l - y_w)} \qquad (21a)$$

In the case of purples, similar elimination of y_s gives the following transformation equations, where y_l' is the trichromatic coefficient (y) of the point on the line joining the extremities of the spectrum locus. This notation serves to distinguish this coördinate of a point on the purple boundary from y_l, the corresponding coördinate of the complementary point on the spectrum locus:

$$p_c = \frac{y_l' - y_w}{y_l - y_w} \cdot \frac{y_l p_e}{y_w + p_e (y_l' - y_w)} \qquad (20b)$$

$$p_e = \frac{y_l - y_w}{y_l' - y_w} \cdot \frac{y_w p_c}{y_l - p_c (y_l - y_w)} \qquad (21b)$$

The above formulae are useful chiefly for the conversion of data that are now expressed in terms of colorimetric purity into values of excitation purity, and occasionally for the reverse procedure.

Note on Chromaticity Diagrams

The chromaticity diagrams appearing on pages 62 to 84 are intended to be used primarily for the determination of dominant wavelength and purity as described on page 59. These charts also provide a means for recording color data in a manner that immediately reveals the relationship of one color to others that have been similarly recorded. Occasionally the large scale of these charts is unnecessary for this purpose, and a small-scale diagram has been included on page 85. The illuminant point has not been indicated on this diagram in order that the diagram may be used for the solution of special problems for which Illuminant C is unsuitable.

For the recording of data, it is preferable to employ duplicates of these charts, rather than the charts themselves. The plates from which the charts were printed have been preserved and, if the demand warrants, duplicates will be available on separate sheets. Information concerning the duplicate charts may be obtained by addressing the Technology Press, Massachusetts Institute of Technology, Cambridge, Massachusetts.

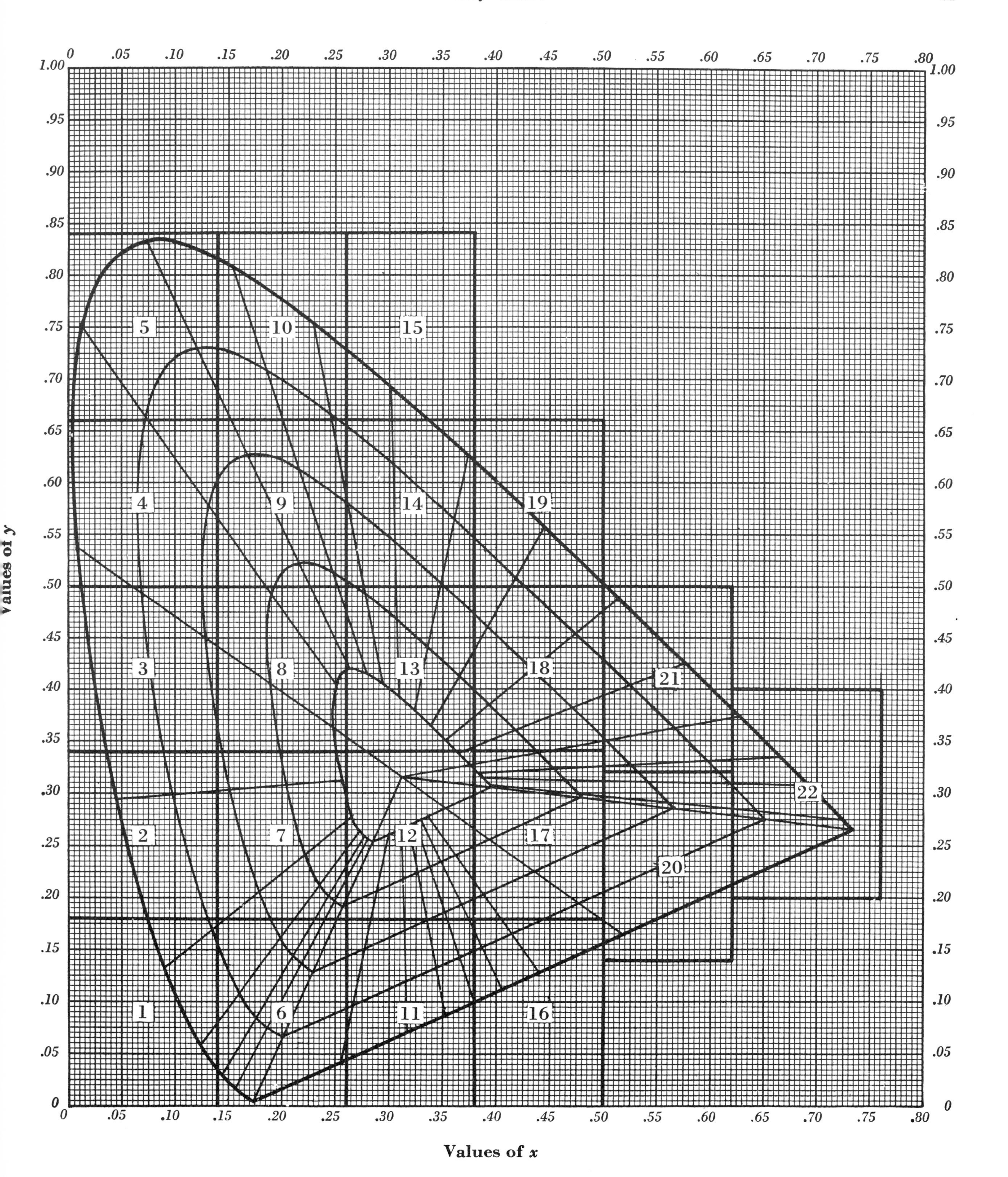
Values of y
Values of x
1
2
3
4
5
6
7
8
9
10
11
12
13
14
15
16
17
18
19
20
21
22

Chart No. 1

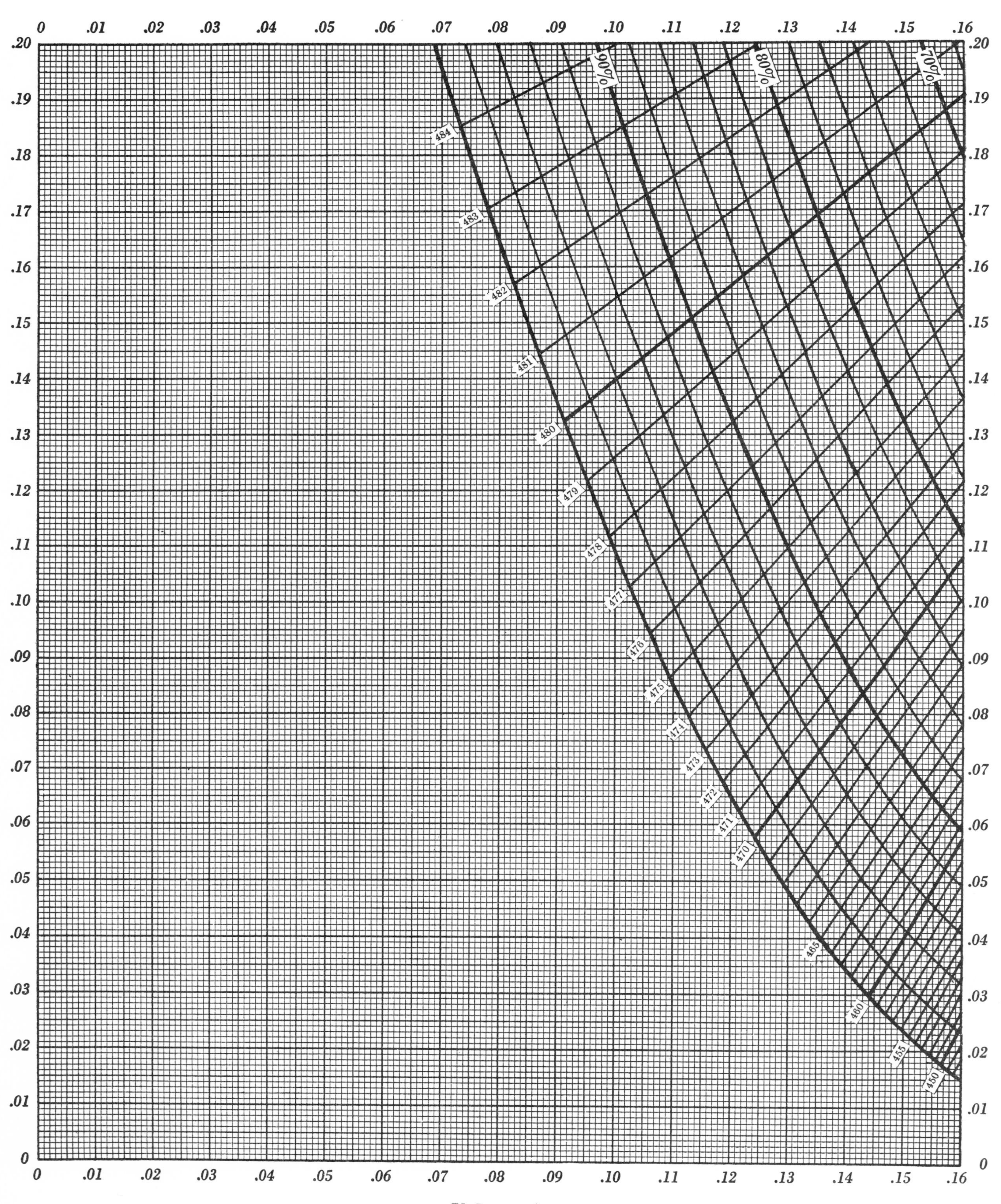

0 .01 .02 .03 .04 .05 .06 .07 .08 .09 .10 .11 .12 .13 .14 .15 .16
.20 .19 .18 .17 .16 .15 .14 .13 .12 .11 .10 .09 .08 .07 .06 .05 .04 .03 .02 .01 0
90%
80%
70%
484
483
482
481
480
479
478
477
476
475
474
473
472
471
470
465
460
455
450
Values of y
Values of x

Chart No. 2

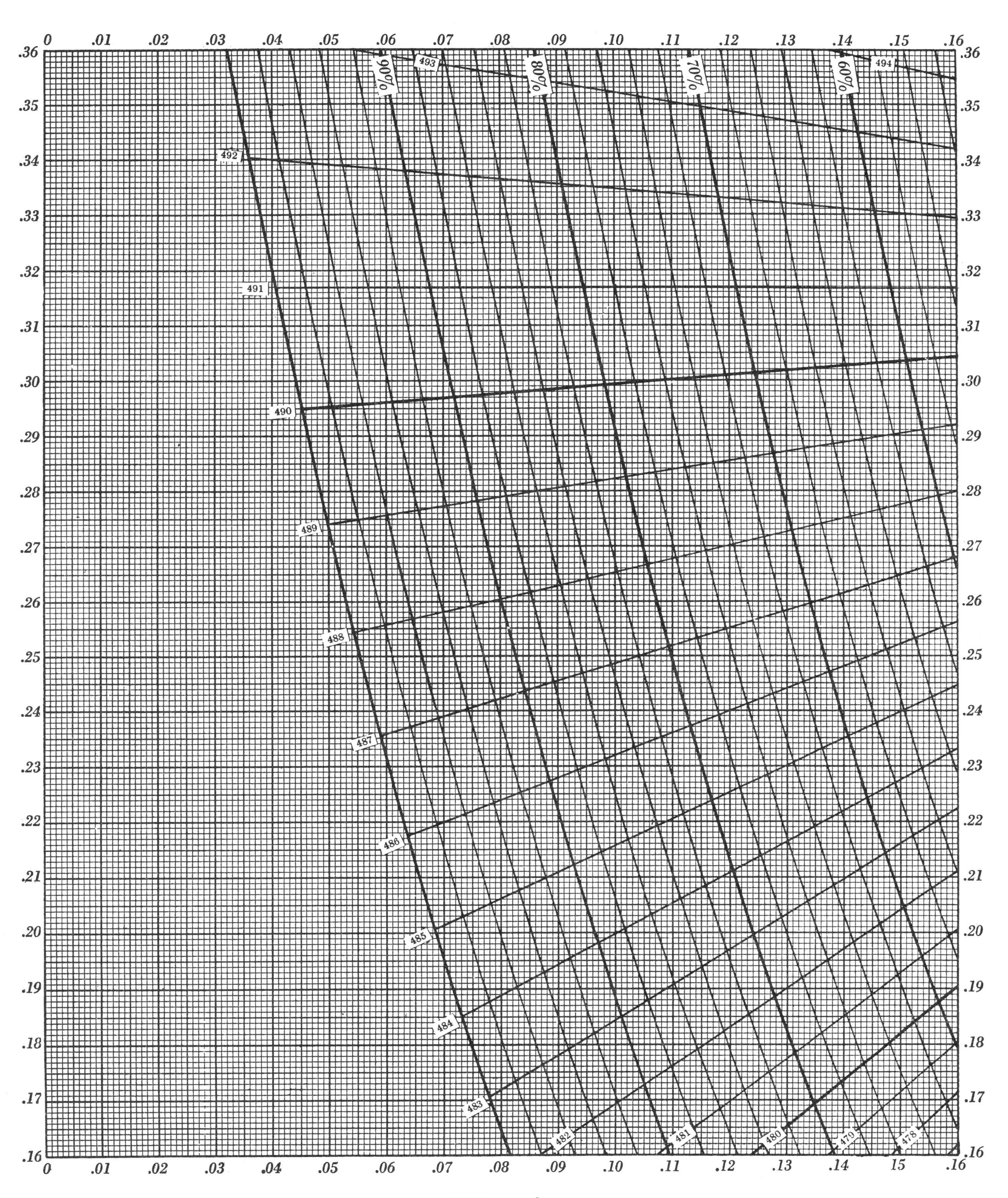

Values of y
Values of x
90%
80%
70%
60%
494
493
492
491
490
489
488
487
486
485
484
483
482
481
480
479
478

Chart No. 3

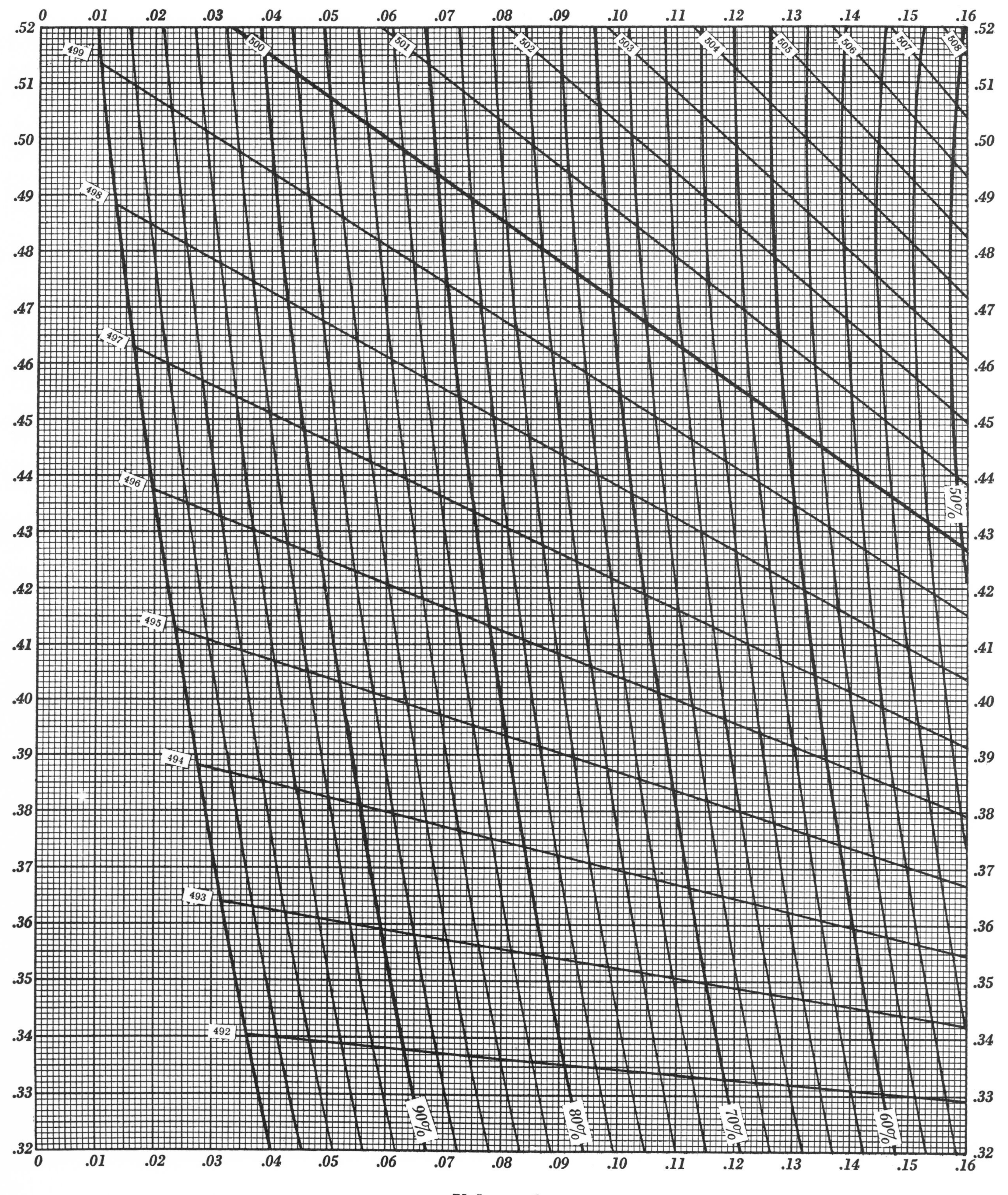
0
.01
.02
.03
.04
.05
.06
.07
.08
.09
.10
.11
.12
.13
.14
.15
.16
.52
.51
.50
.49
.48
.47
.46
.45
.44
.43
.42
.41
.40
.39
.38
.37
.36
.35
.34
.33
.32
499
498
497
496
495
494
493
492
500
501
502
503
504
505
506
507
508
50%
60%
70%
80%
90%
Values of y
Values of x

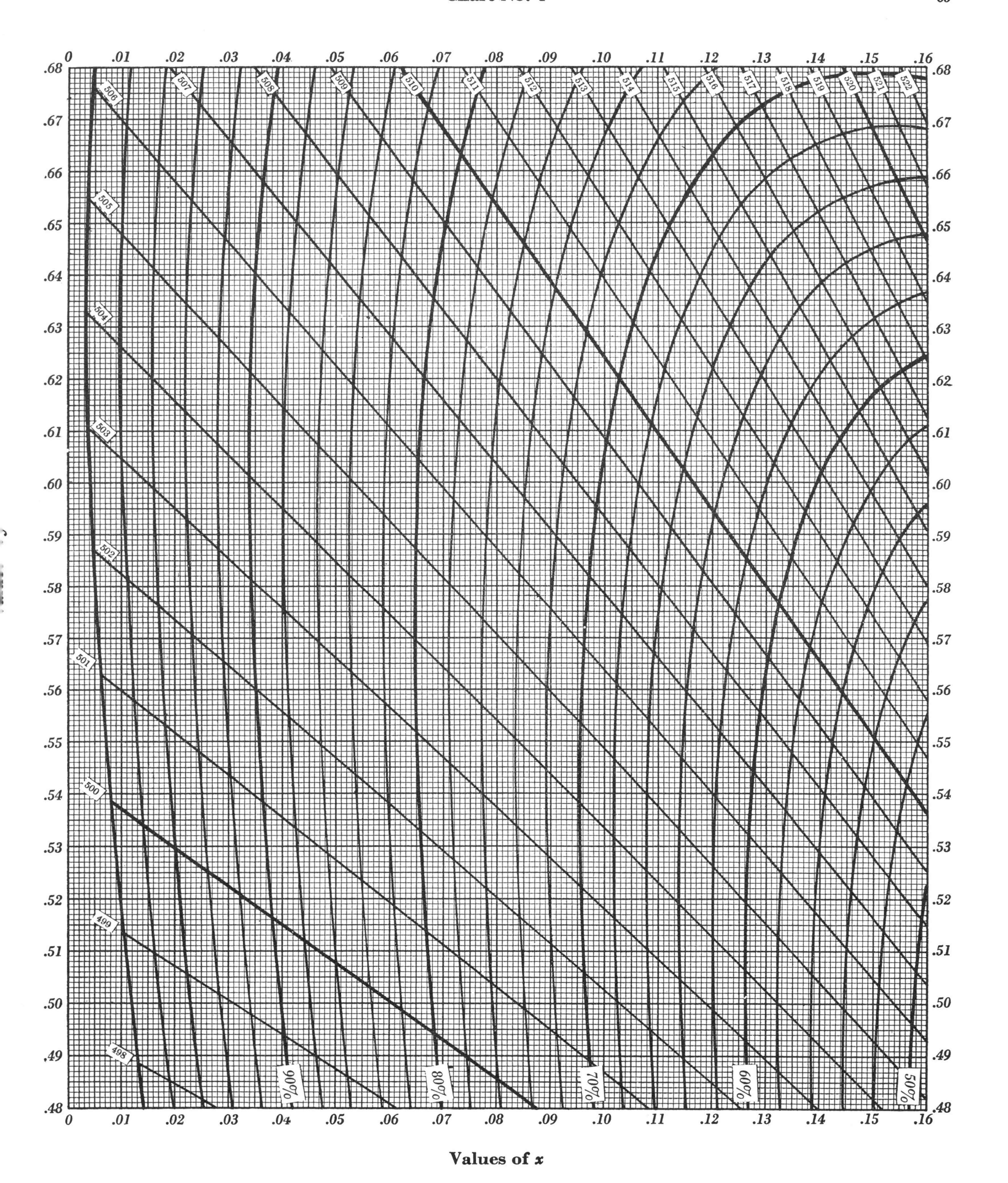
0
.01
.02
.03
.04
.05
.06
.07
.08
.09
.10
.11
.12
.13
.14
.15
.16
.68
.67
.66
.65
.64
.63
.62
.61
.60
.59
.58
.57
.56
.55
.54
.53
.52
.51
.50
.49
.48
498
499
500
501
502
503
504
505
506
507
508
509
510
511
512
513
514
515
516
517
518
519
520
521
522
90%
80%
70%
60%
50%
Values of x

Chart No. 5

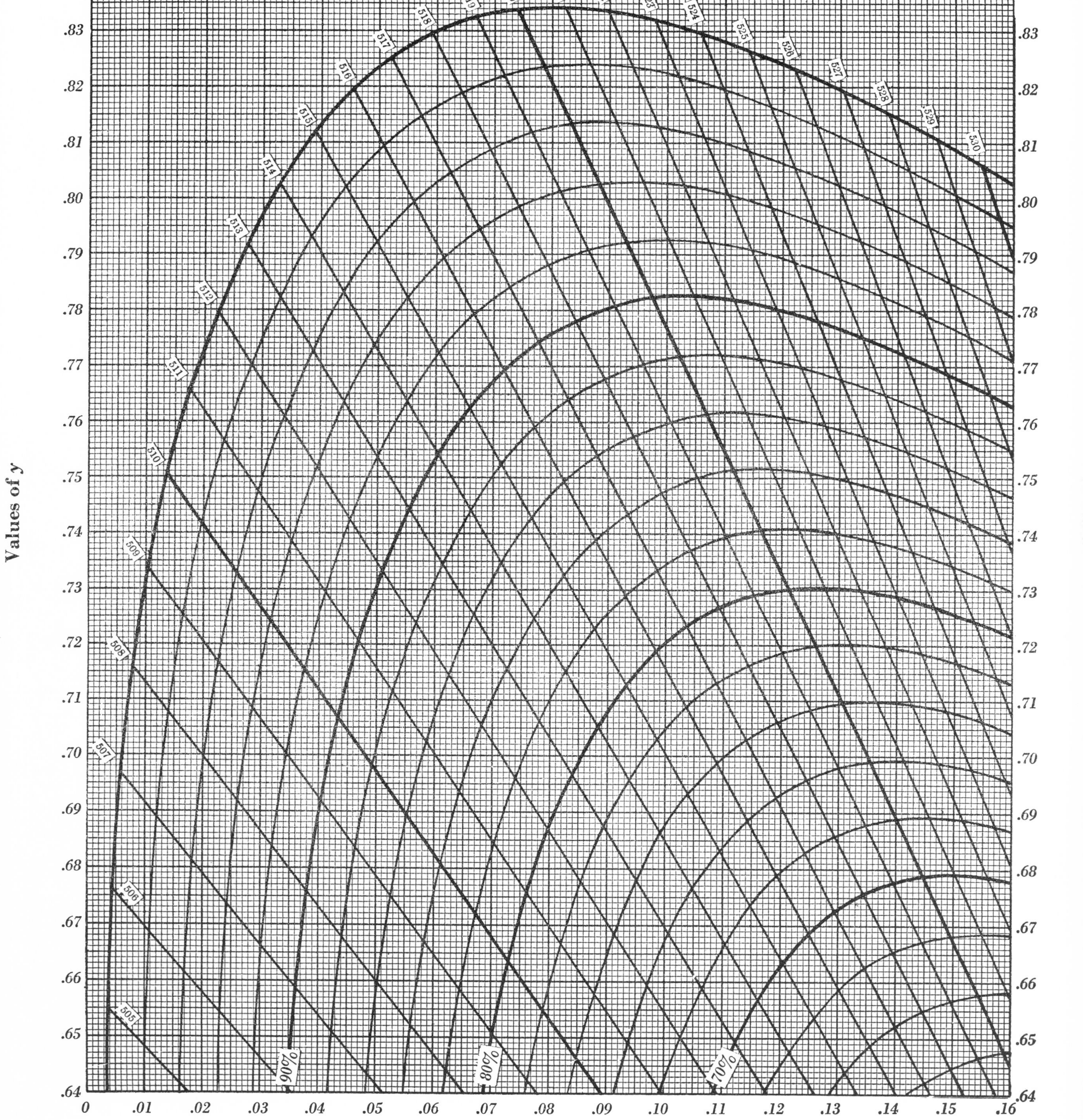

0
.01
.02
.03
.04
.05
.06
.07
.08
.09
.10
.11
.12
.13
.14
.15
.16
.84
.83
.82
.81
.80
.79
.78
.77
.76
.75
.74
.73
.72
.71
.70
.69
.68
.67
.66
.65
.64
505
506
507
508
509
510
511
512
513
514
515
516
517
518
519
520
521
522
523
524
525
526
527
528
529
530
90%
80%
70%
Values of y
Values of x

Chart No. 6

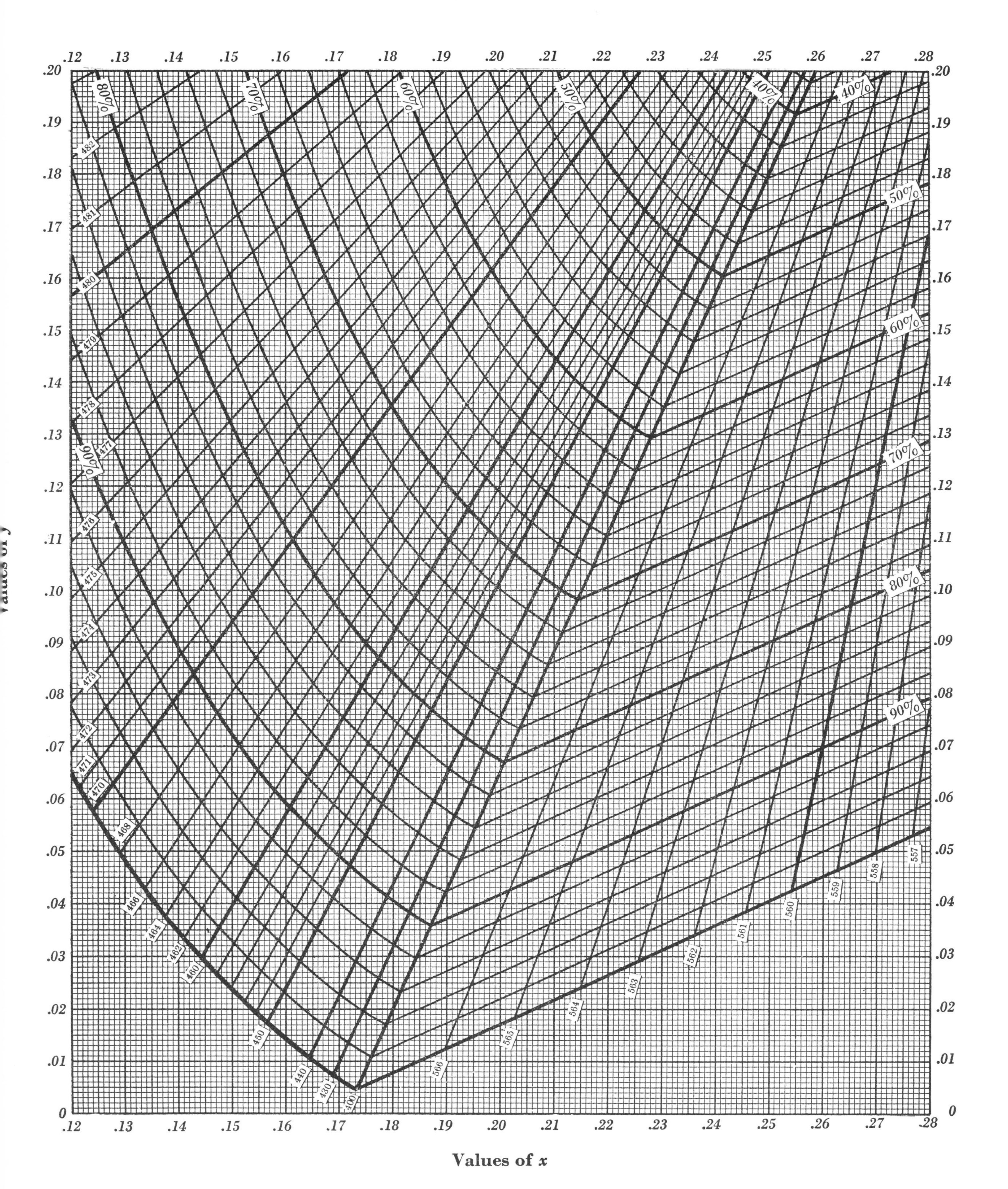

Chart No. 7

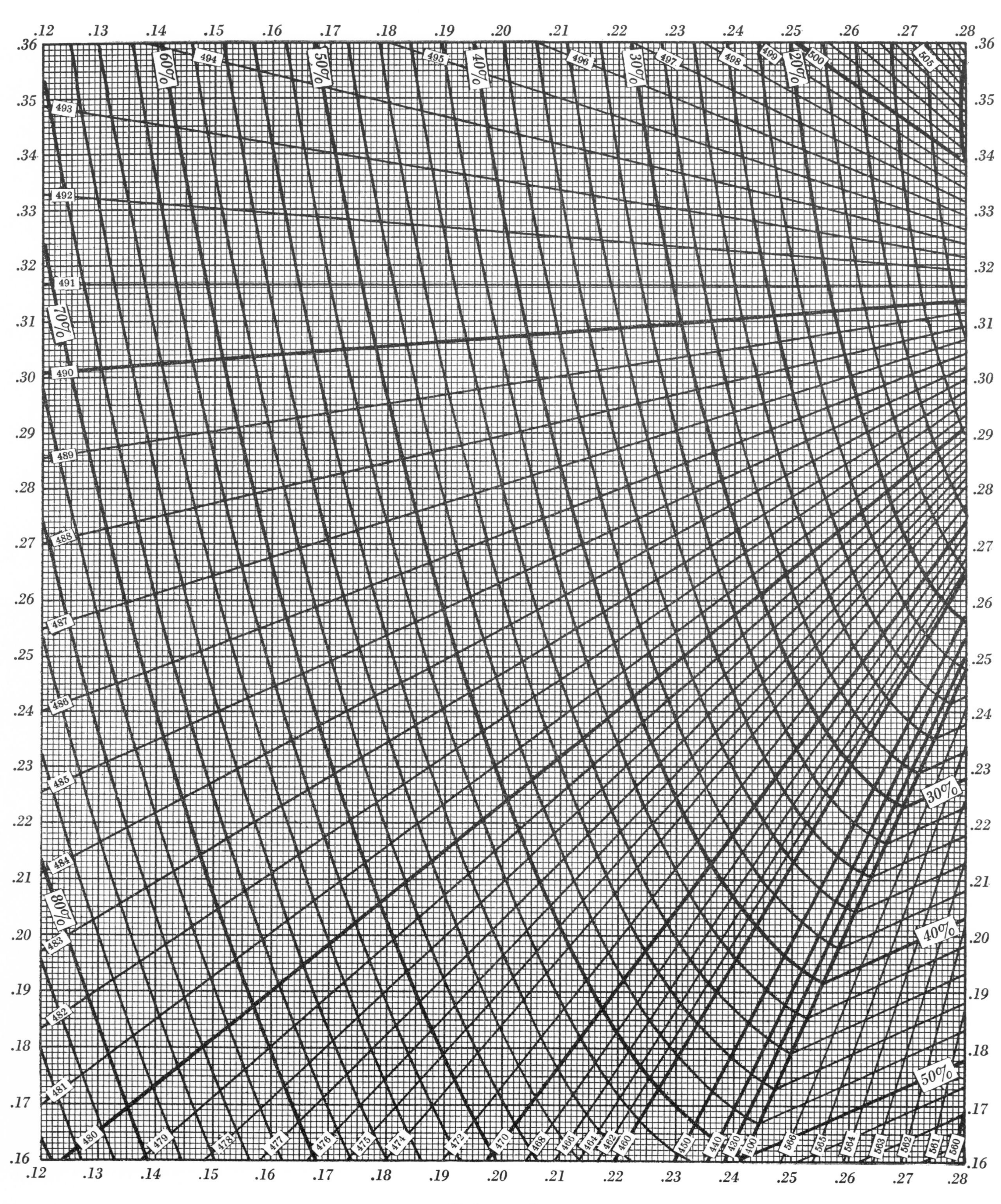
Values of y
Values of x
.12
.13
.14
.15
.16
.17
.18
.19
.20
.21
.22
.23
.24
.25
.26
.27
.28
.36
.35
.34
.33
.32
.31
.30
.29
.28
.27
.26
.25
.24
.23
.22
.21
.20
.19
.18
.17
.16
60%
50%
40%
30%
20%
70%
80%
30%
40%
50%

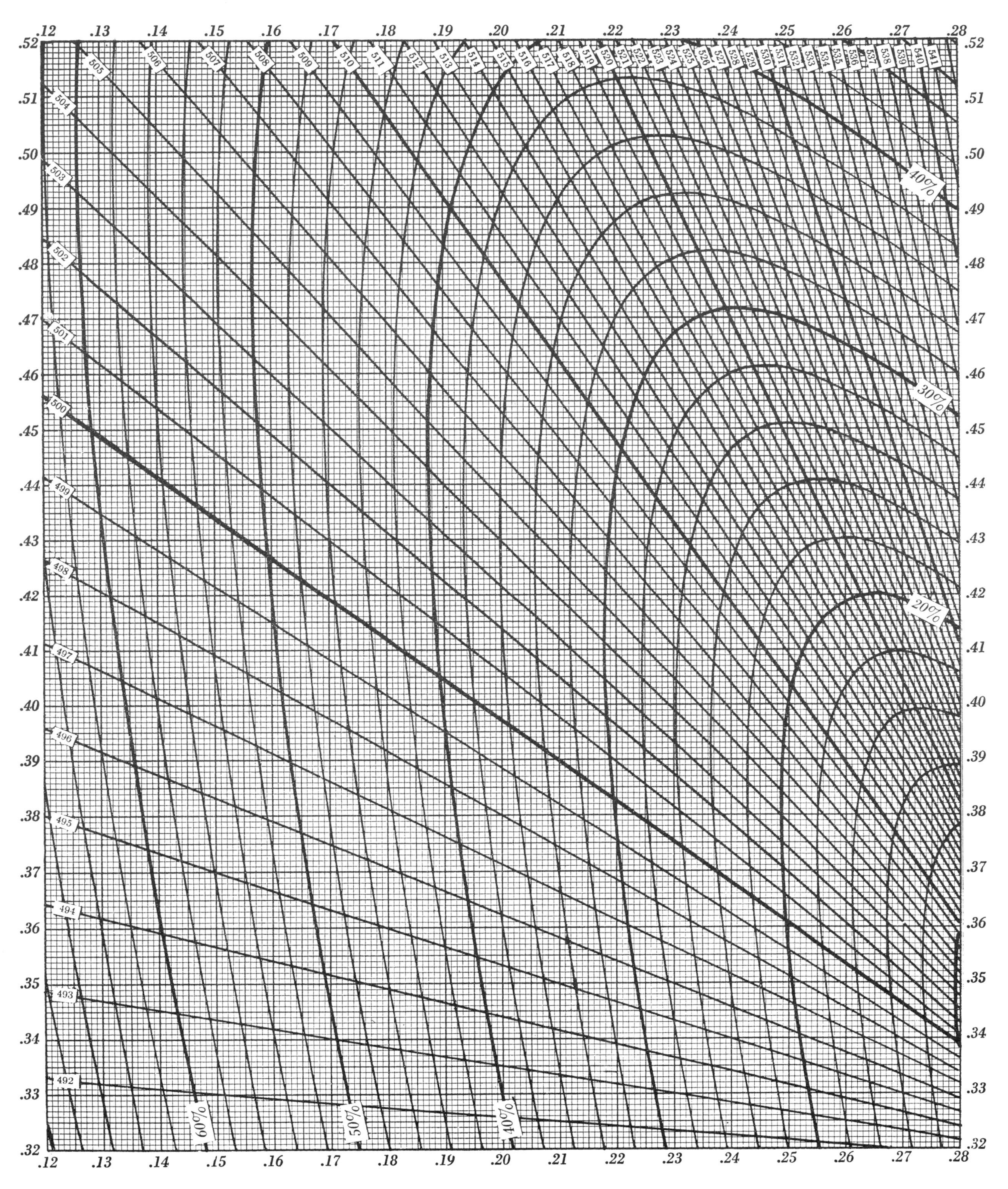

Values of y
Values of x
.12 .13 .14 .15 .16 .17 .18 .19 .20 .21 .22 .23 .24 .25 .26 .27 .28
.32 .33 .34 .35 .36 .37 .38 .39 .40 .41 .42 .43 .44 .45 .46 .47 .48 .49 .50 .51 .52
492 493 494 495 496 497 498 499 500 501 502 503 504 505 506 507 508 509 510 511 512 513 514 515 516 517 518 519 520 521 522 523 524 525 526 527 528 529 530 531 532 533 534 535 536 537 538 539 540 541
60% 50% 40% 40% 30% 20%

Chart No. 9

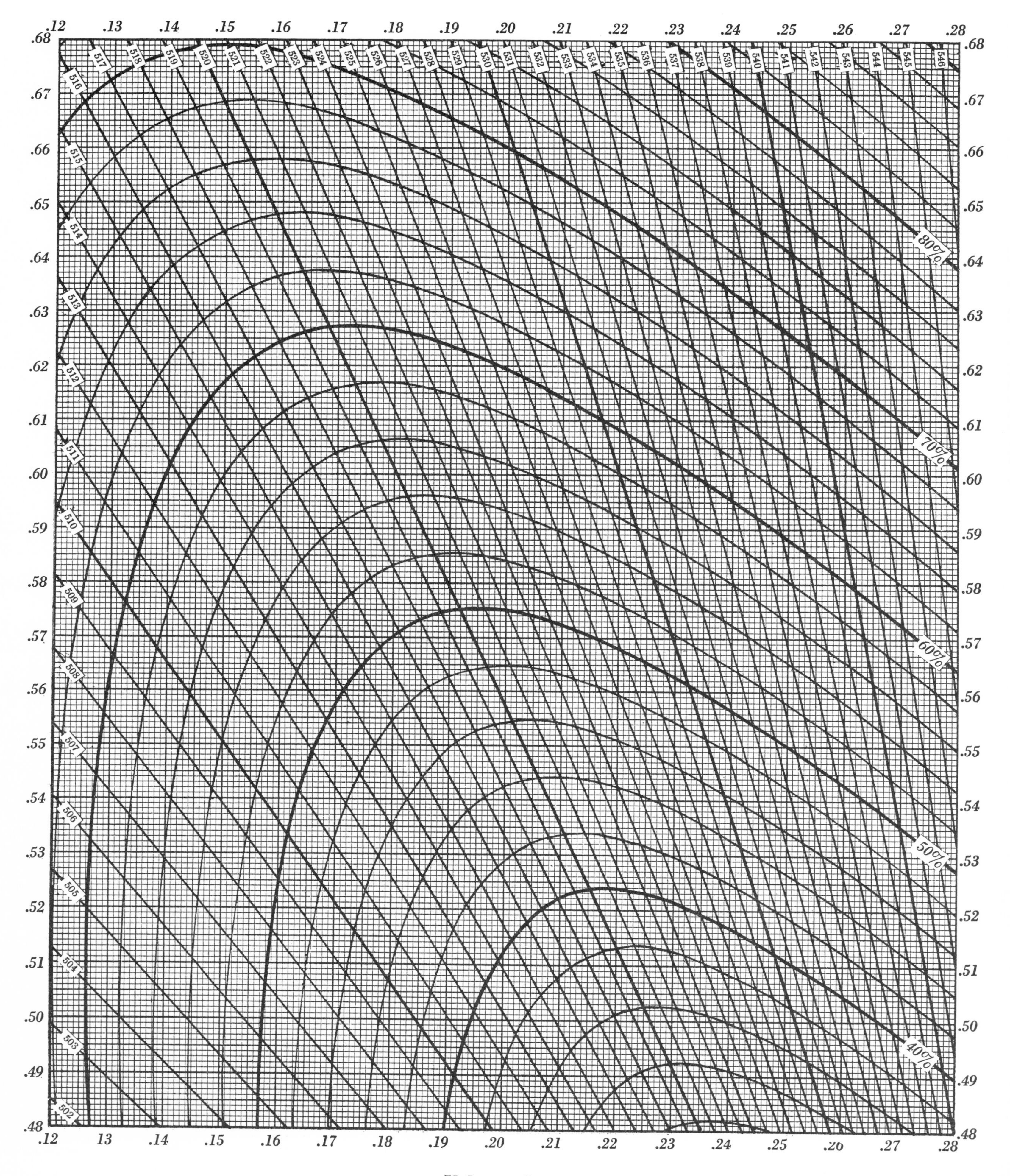

Values of y
Values of x
.12 .13 .14 .15 .16 .17 .18 .19 .20 .21 .22 .23 .24 .25 .26 .27 .28
.48 .49 .50 .51 .52 .53 .54 .55 .56 .57 .58 .59 .60 .61 .62 .63 .64 .65 .66 .67 .68
502 503 504 505 506 507 508 509 510 511 512 513 514 515 516 517 518 519 520 521 522 523 524 525 526 527 528 529 530 531 532 533 534 535 536 537 538 539 540 541 542 543 544 545 546
40%
50%
60%
70%
80%

Chart No. 10

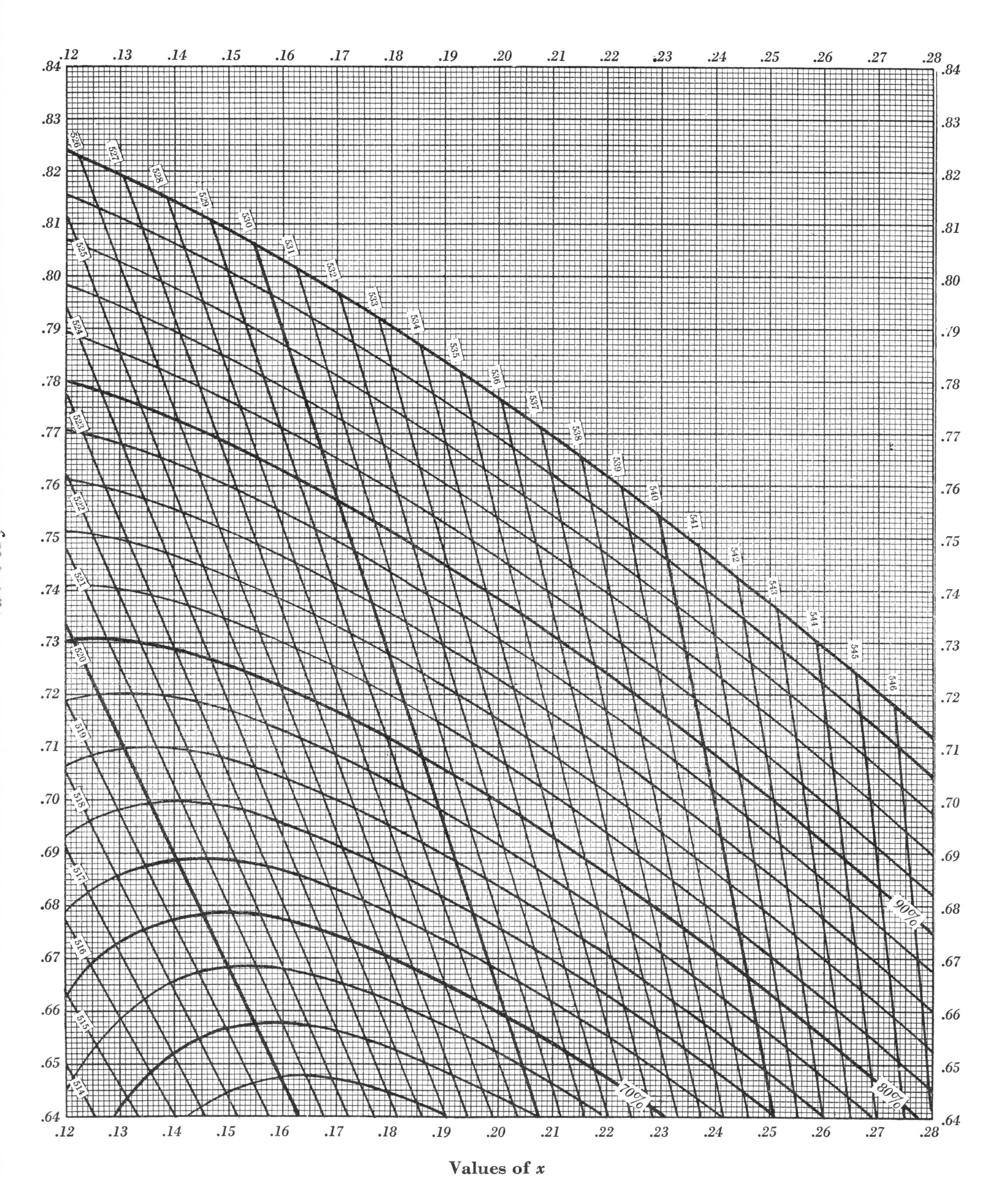
.12
.13
.14
.15
.16
.17
.18
.19
.20
.21
.22
.23
.24
.25
.26
.27
.28
.84
.83
.82
.81
.80
.79
.78
.77
.76
.75
.74
.73
.72
.71
.70
.69
.68
.67
.66
.65
.64
514
515
516
517
518
519
520
521
522
523
524
525
526
527
528
529
530
531
532
533
534
535
536
537
538
539
540
541
542
543
544
545
546
70%
80%
90%
Values of x

Chart No. 11

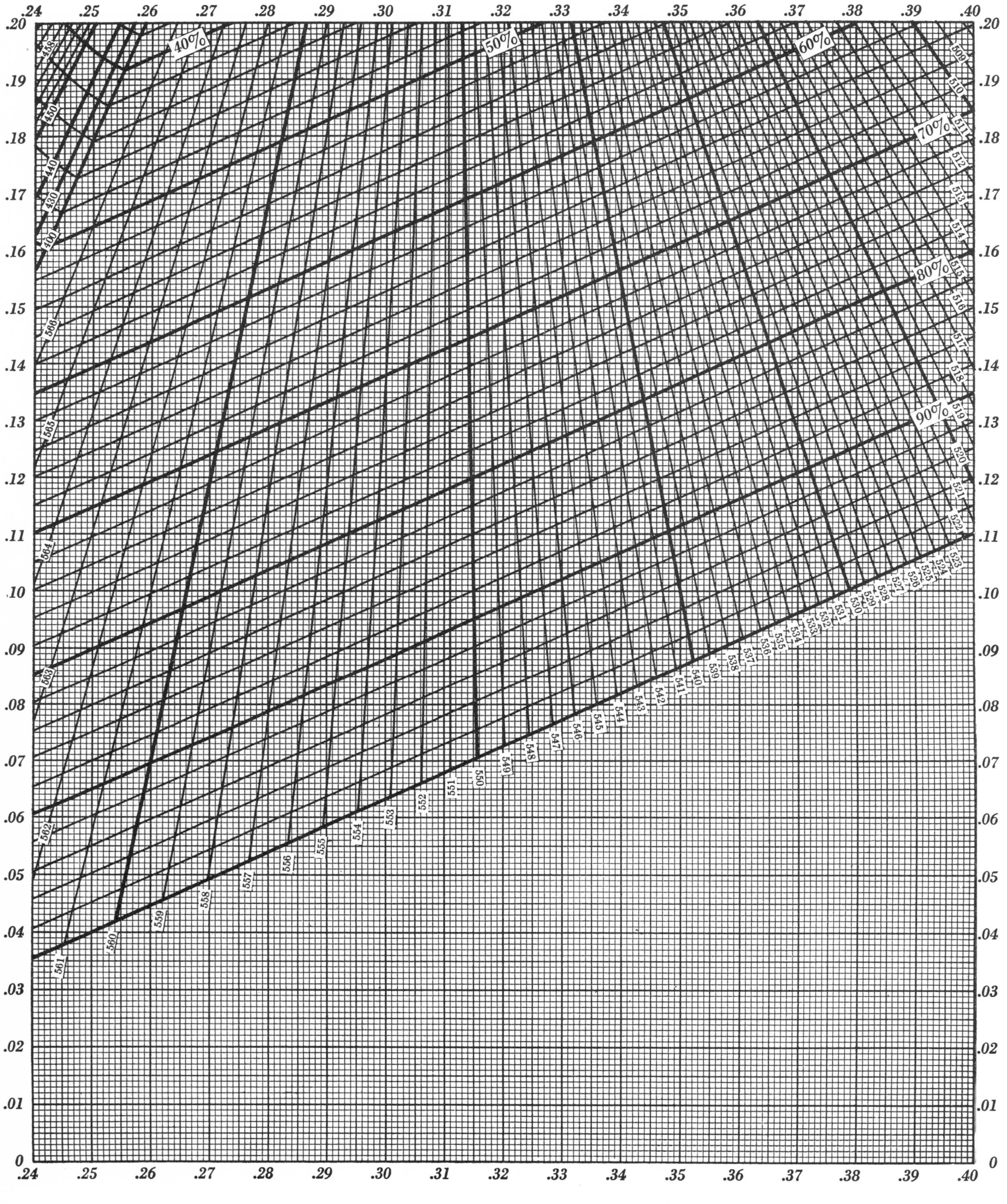

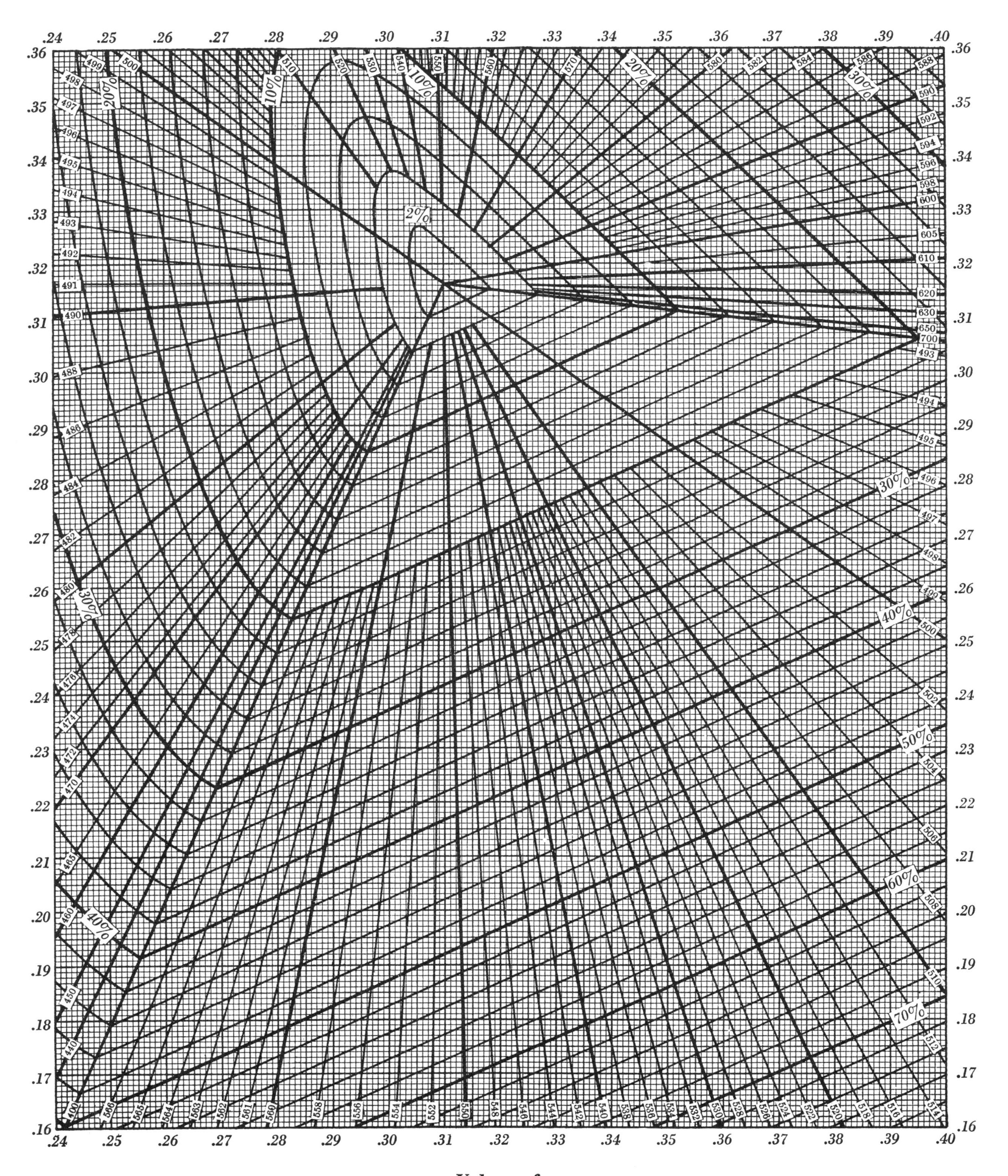
Values of x
Values of y
2%
10%
20%
30%
40%
50%
60%
70%

Chart No. 12A

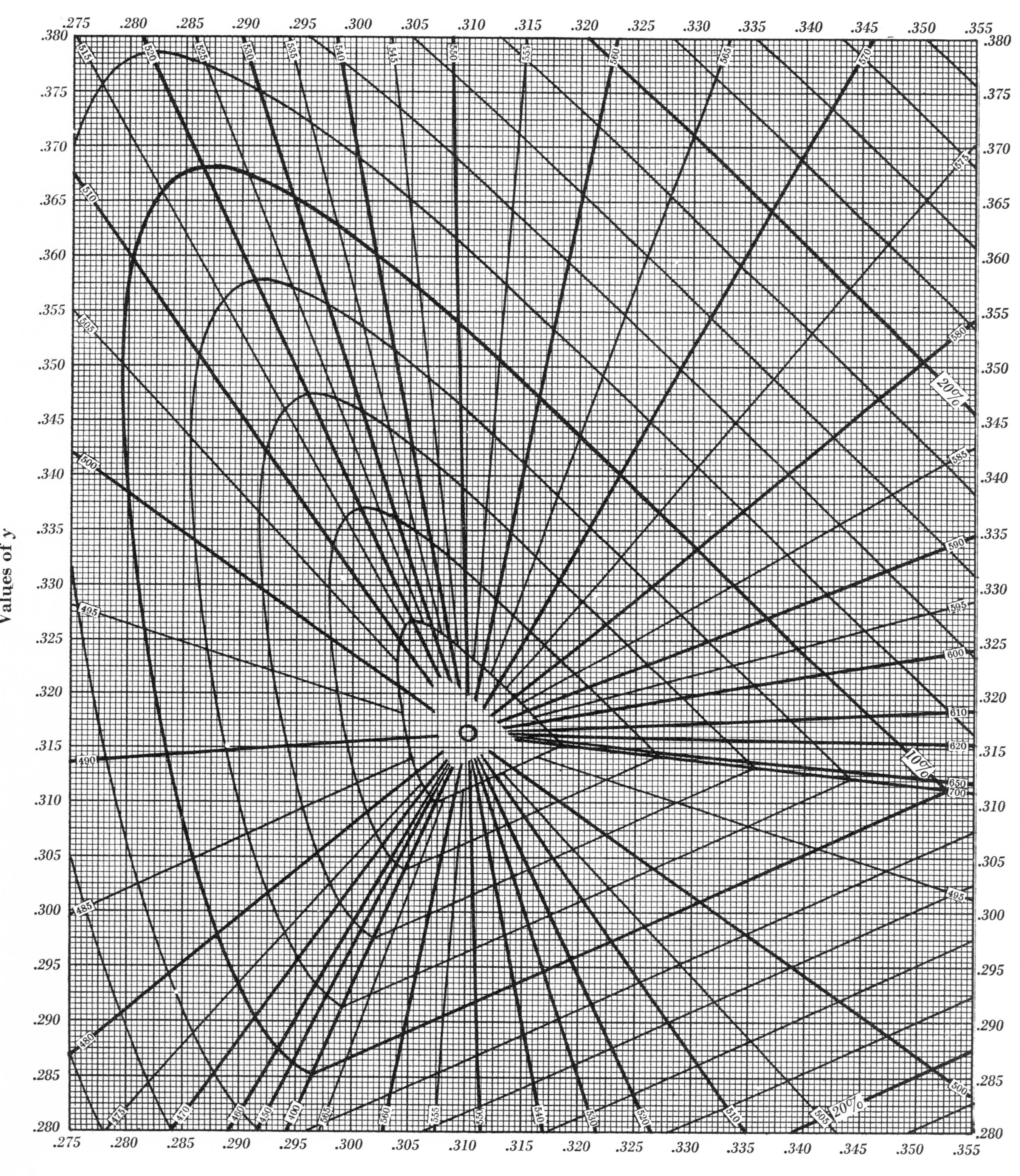
Values of y
Values of x
.275
.280
.285
.290
.295
.300
.305
.310
.315
.320
.325
.330
.335
.340
.345
.350
.355
.380
.375
.370
.365
.360
.355
.350
.345
.340
.335
.330
.325
.320
.315
.310
.305
.300
.295
.290
.285
.280
20%
10%
20%

Chart No. 13

Chart No. 14

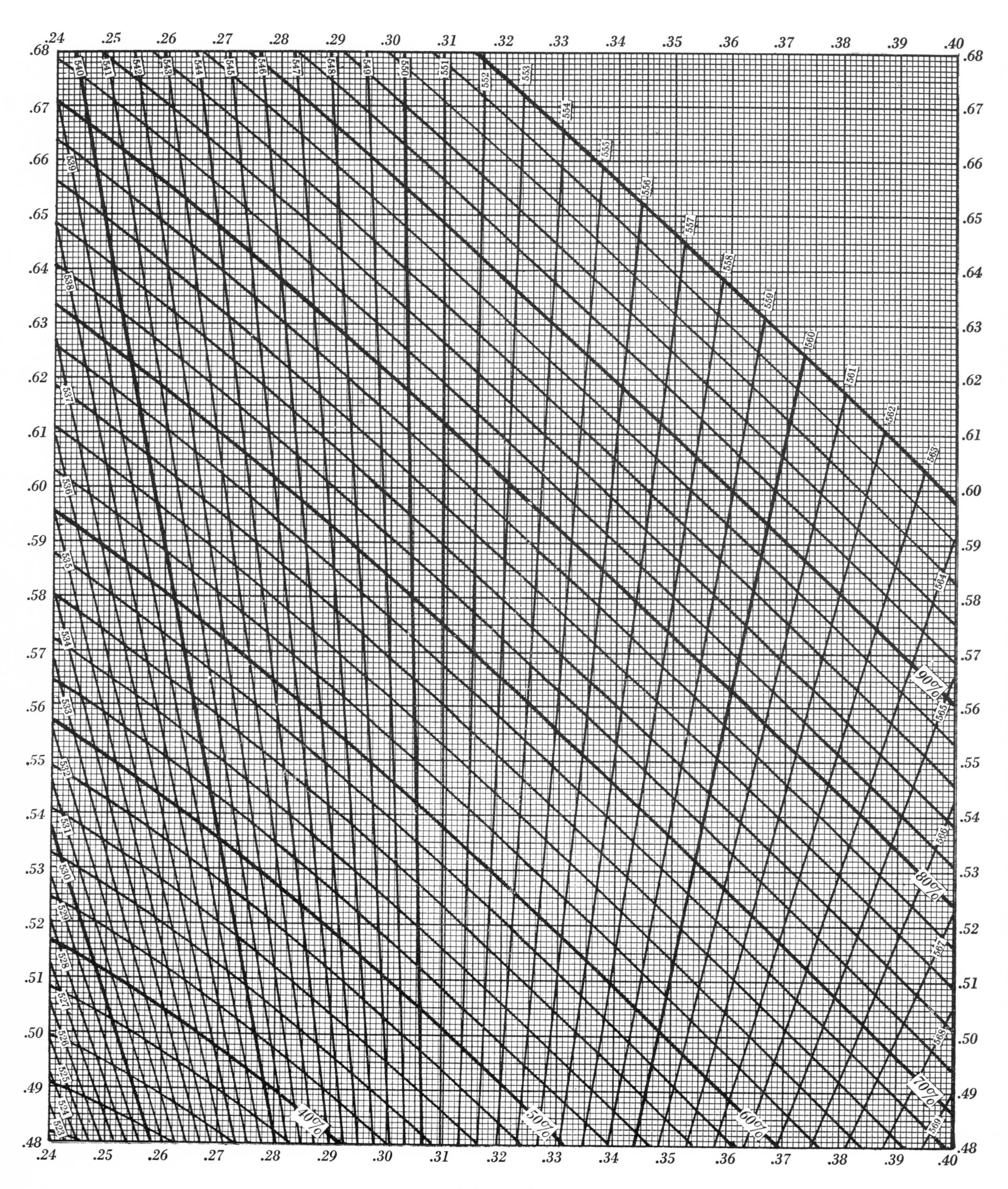
Values of y
Values of x
.24
.25
.26
.27
.28
.29
.30
.31
.32
.33
.34
.35
.36
.37
.38
.39
.40
.48
.49
.50
.51
.52
.53
.54
.55
.56
.57
.58
.59
.60
.61
.62
.63
.64
.65
.66
.67
.68
40%
50%
60%
70%
80%
90%

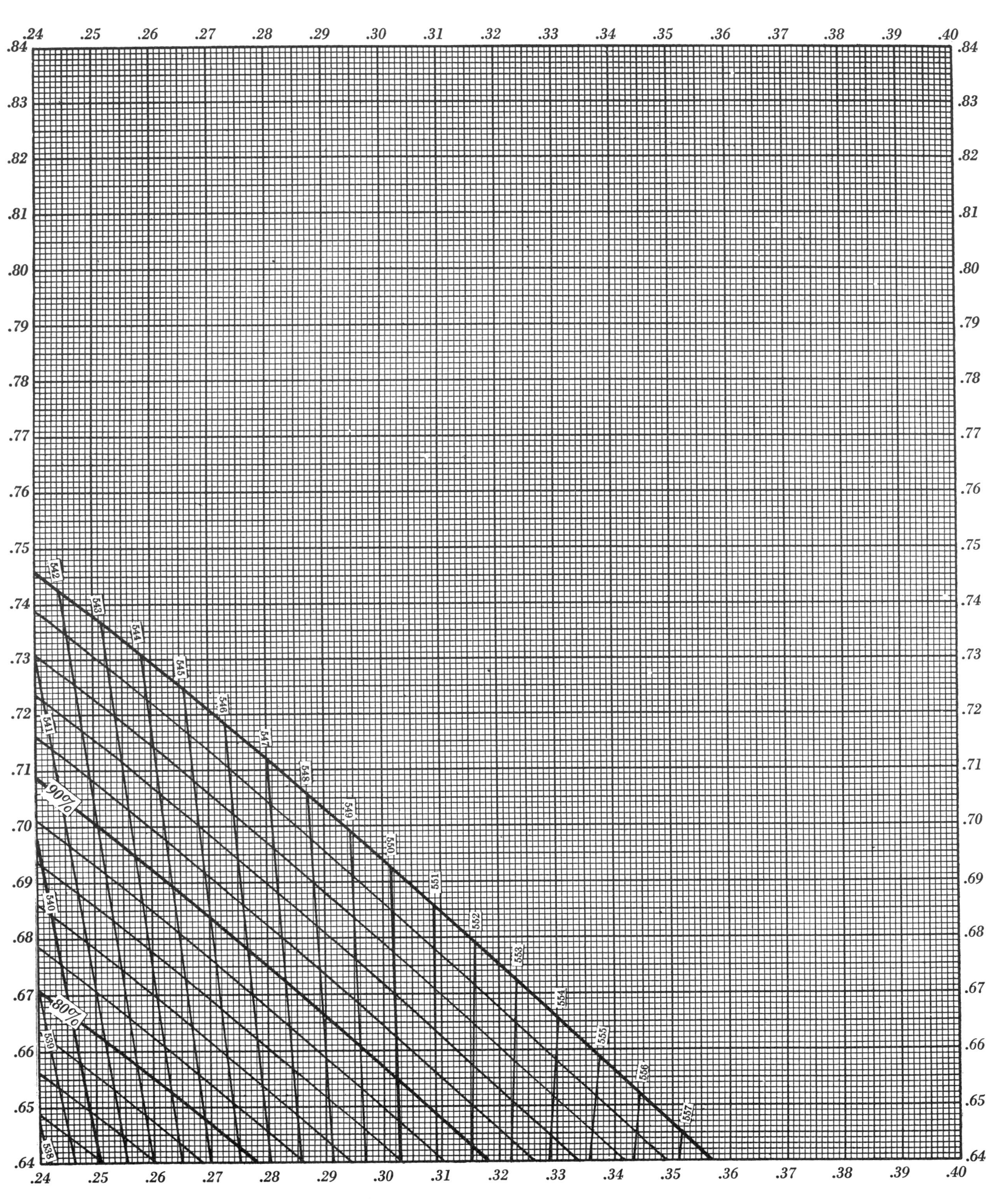
.24
.25
.26
.27
.28
.29
.30
.31
.32
.33
.34
.35
.36
.37
.38
.39
.40
.84
.83
.82
.81
.80
.79
.78
.77
.76
.75
.74
.73
.72
.71
.70
.69
.68
.67
.66
.65
.64
538
539
540
541
542
543
544
545
546
547
548
549
550
551
552
553
554
555
556
557
90%
80%
Values of x

Chart No. 16

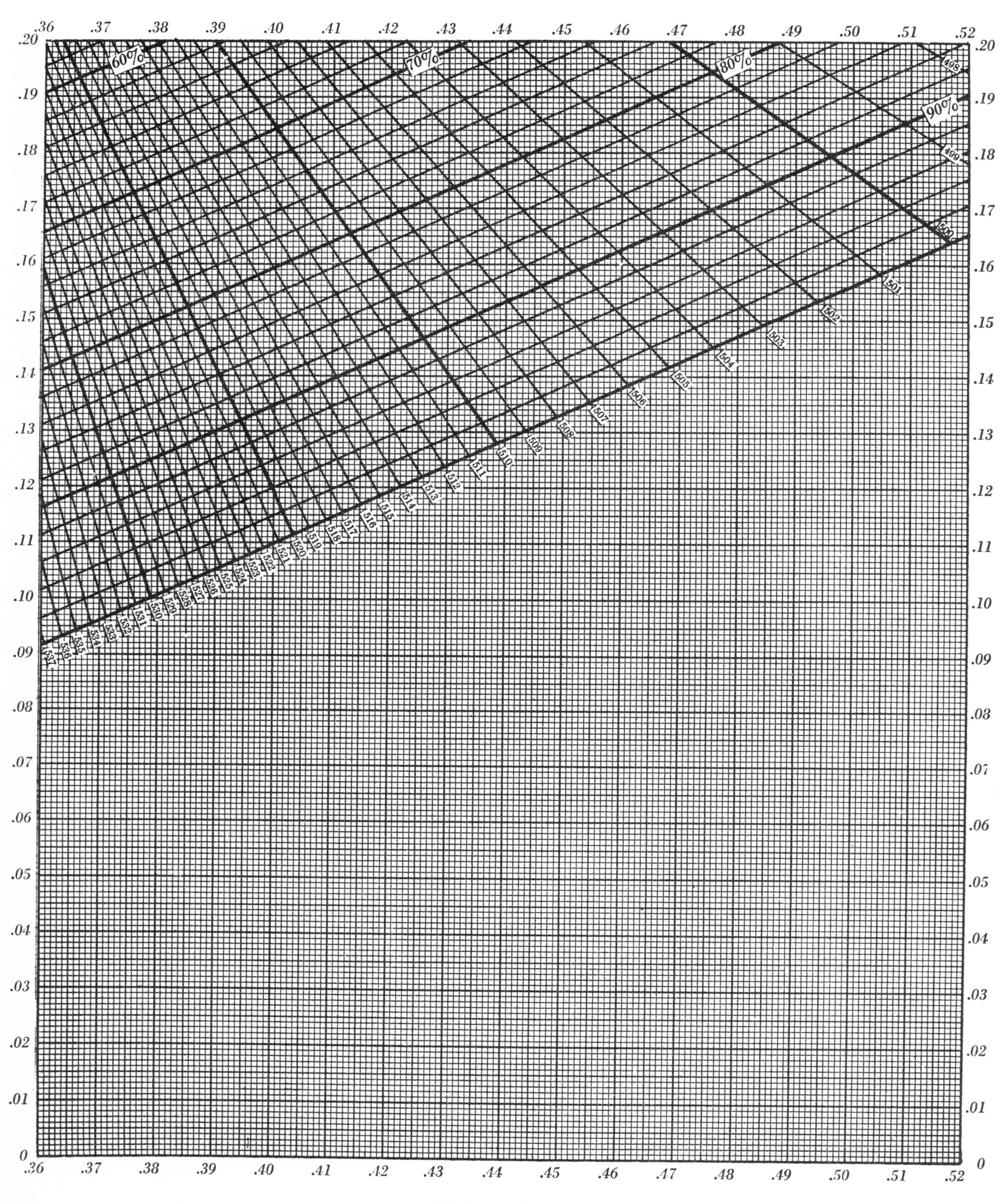
Values of y
Values of x
60%
70%
80%
90%
.36
.37
.38
.39
.40
.41
.42
.43
.44
.45
.46
.47
.48
.49
.50
.51
.52
0
.01
.02
.03
.04
.05
.06
.07
.08
.09
.10
.11
.12
.13
.14
.15
.16
.17
.18
.19
.20
498
499
500
501
502
503
504
505
506
507
508
509
510
511
512
513
514
515
516
517
518
519
520
521
522
523
524
525
526
527
528
529
530
531
532
533
534
535
536
537

Chart No. 17

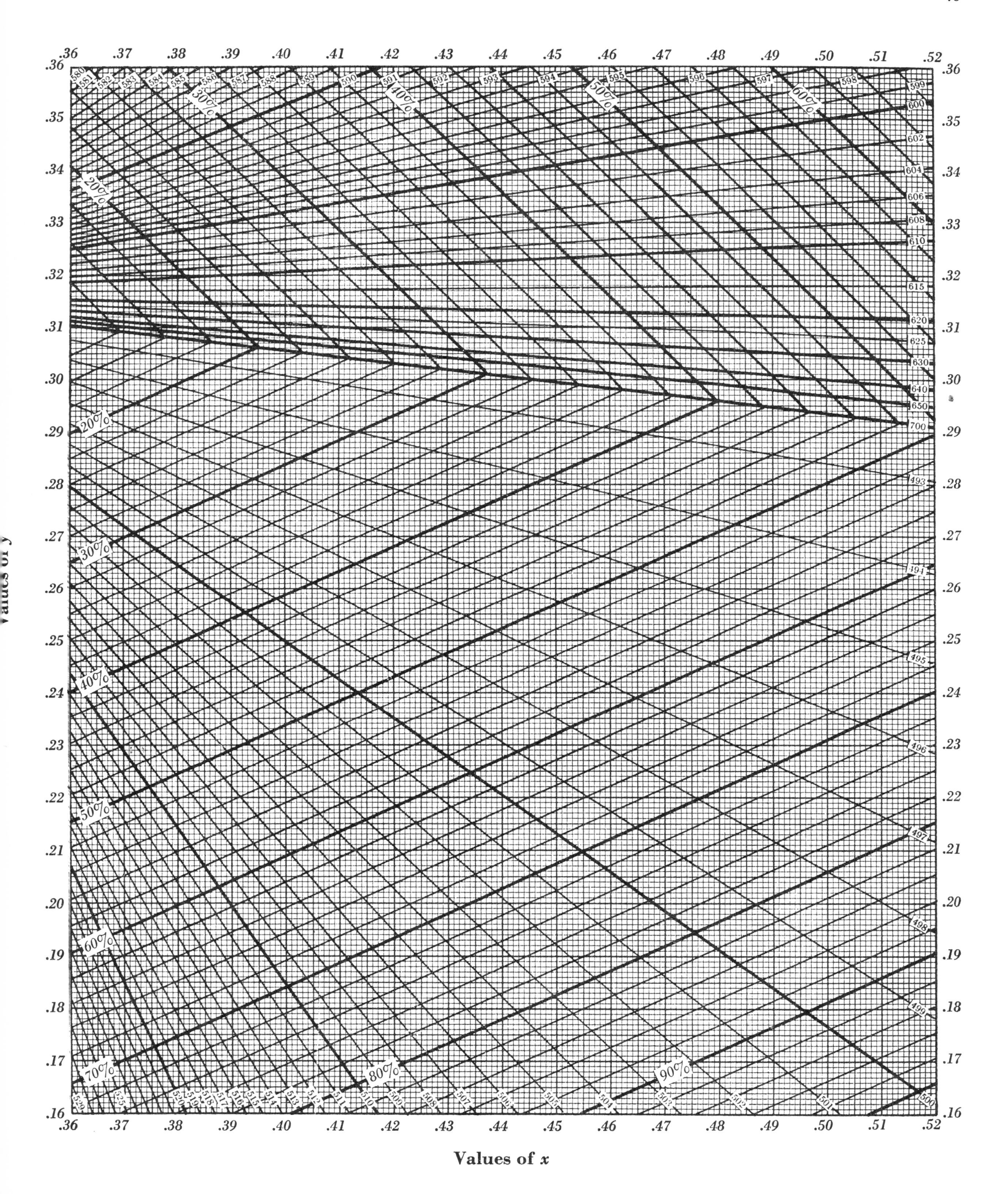

Chart No. 18

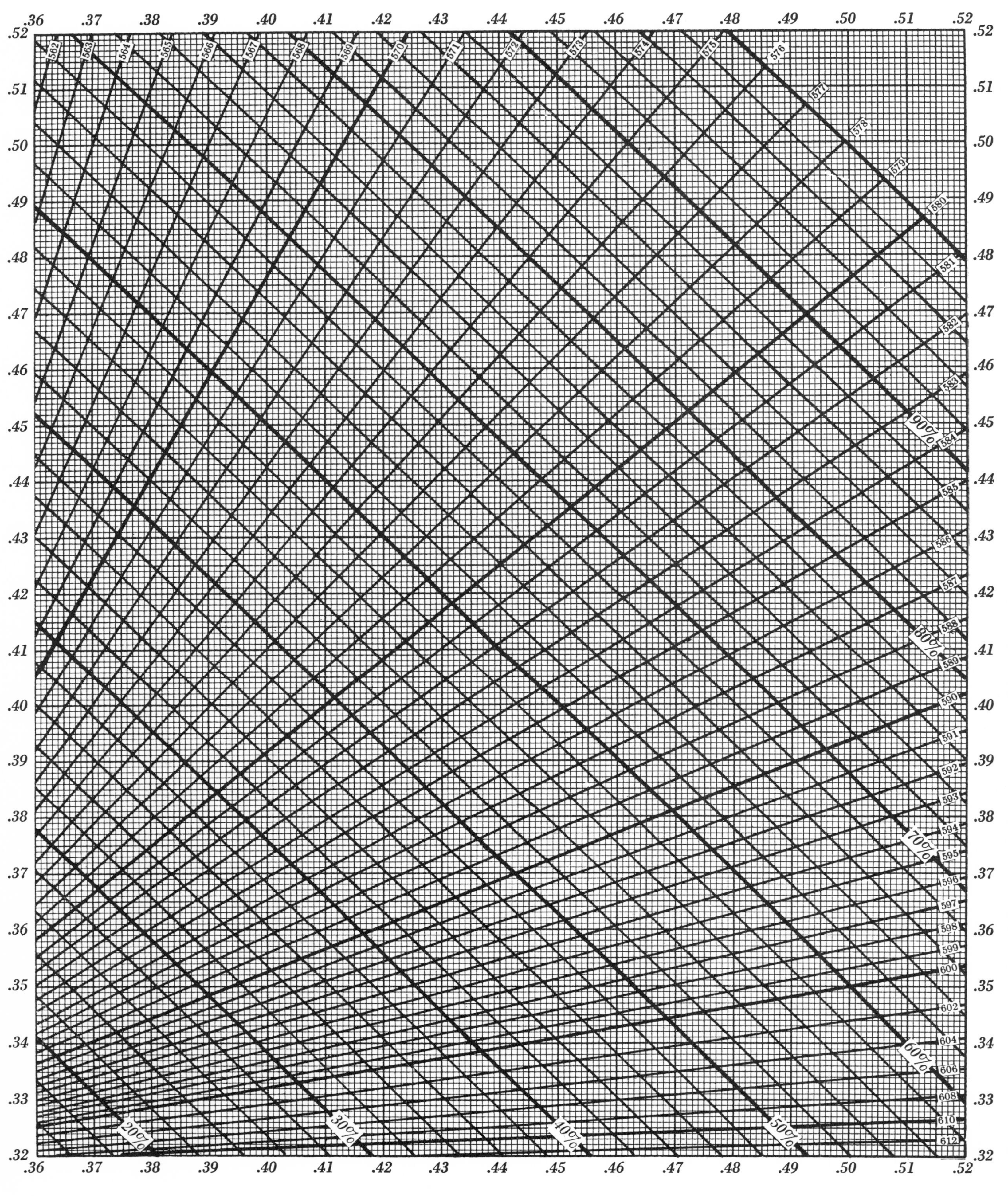

.36 .37 .38 .39 .40 .41 .42 .43 .44 .45 .46 .47 .48 .49 .50 .51 .52
.52 .51 .50 .49 .48 .47 .46 .45 .44 .43 .42 .41 .40 .39 .38 .37 .36 .35 .34 .33 .32
562 563 564 565 566 567 568 569 570 571 572 573 574 575 576 577 578 579 580 581 582 583 584 585 586 587 588 589 590 591 592 593 594 595 596 597 598 599 600 602 604 606 608 610 612
20% 30% 40% 50% 60% 70% 80% 90%
Values of y
Values of x

Chart No. 19

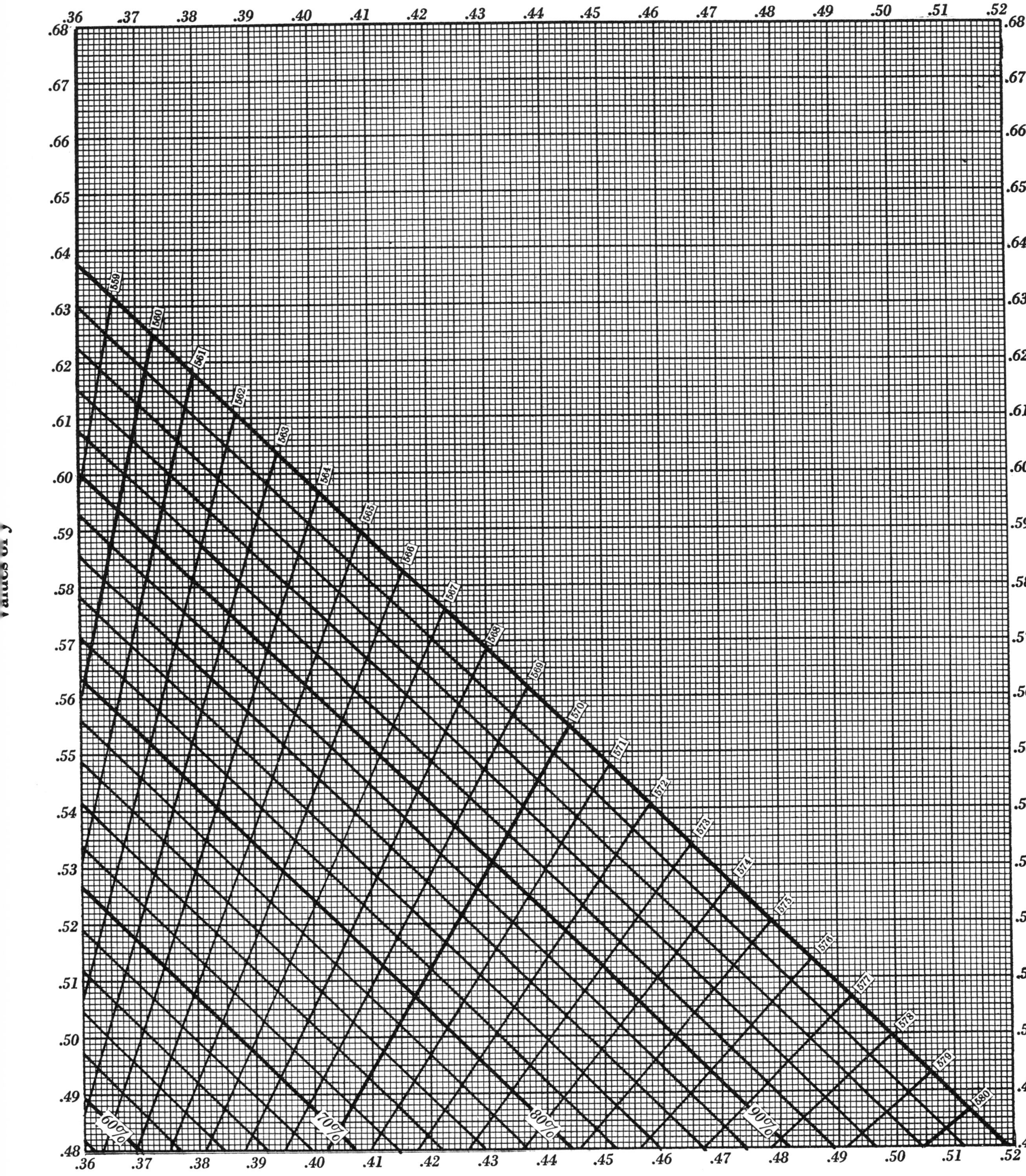

.36 .37 .38 .39 .40 .41 .42 .43 .44 .45 .46 .47 .48 .49 .50 .51 .52

.68 .67 .66 .65 .64 .63 .62 .61 .60 .59 .58 .57 .56 .55 .54 .53 .52 .51 .50 .49 .48

559 560 561 562 563 564 565 566 567 568 569 570 571 572 573 574 575 576 577 578 579 580

60% 70% 80% 90%

.36 .37 .38 .39 .40 .41 .42 .43 .44 .45 .46 .47 .48 .49 .50 .51 .52

Values of x

Values of y

Chart No. 20

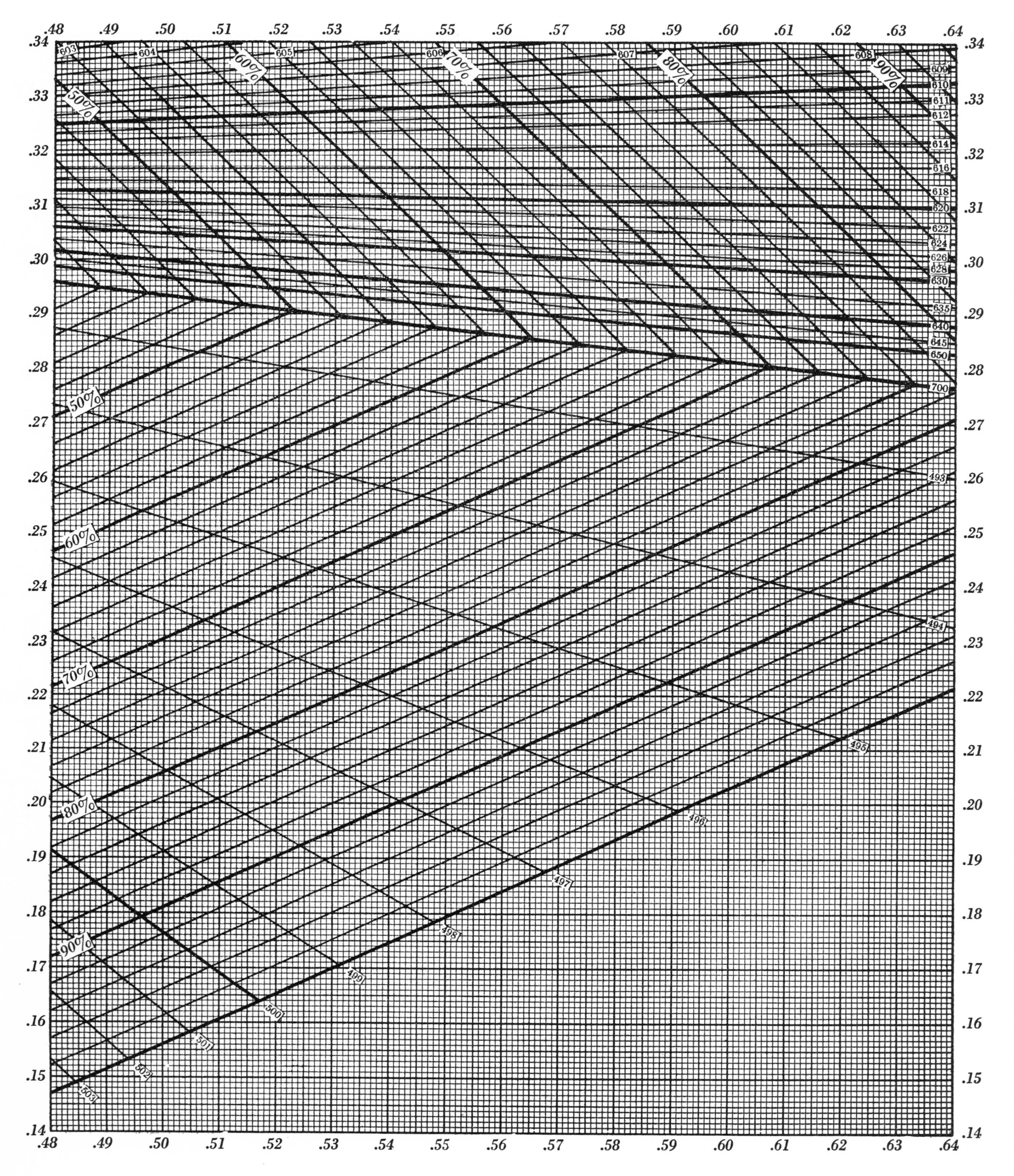

Values of y
Values of x
.48 .49 .50 .51 .52 .53 .54 .55 .56 .57 .58 .59 .60 .61 .62 .63 .64
.34 .33 .32 .31 .30 .29 .28 .27 .26 .25 .24 .23 .22 .21 .20 .19 .18 .17 .16 .15 .14
50%
60%
70%
80%
90%
603
604
605
606
607
608
609
610
611
612
614
616
618
620
622
624
626
628
630
635
640
645
650
700
493
494
495
496
497
498
499
500
501
502
503

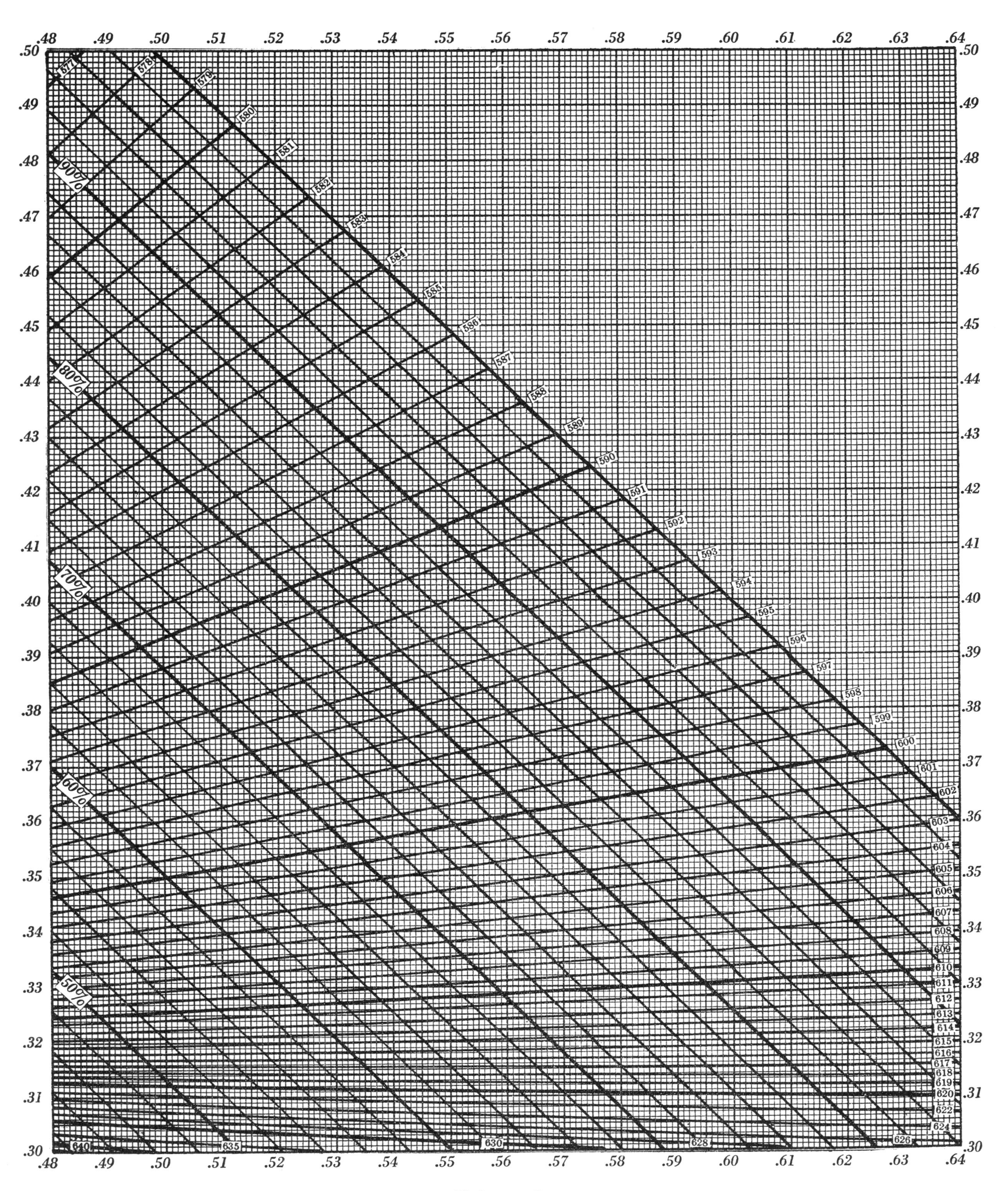
.48 .49 .50 .51 .52 .53 .54 .55 .56 .57 .58 .59 .60 .61 .62 .63 .64
.50 .49 .48 .47 .46 .45 .44 .43 .42 .41 .40 .39 .38 .37 .36 .35 .34 .33 .32 .31 .30
577 578 579 580 581 582 583 584 585 586 587 588 589 590 591 592 593 594 595 596 597 598 599 600 601 602 603 604 605 606 607 608 609 610 611 612 613 614 615 616 617 618 619 620 622 624 626 628 630 635 640
90% 80% 70% 60% 50%
Values of y
Values of x

Chart No. 22

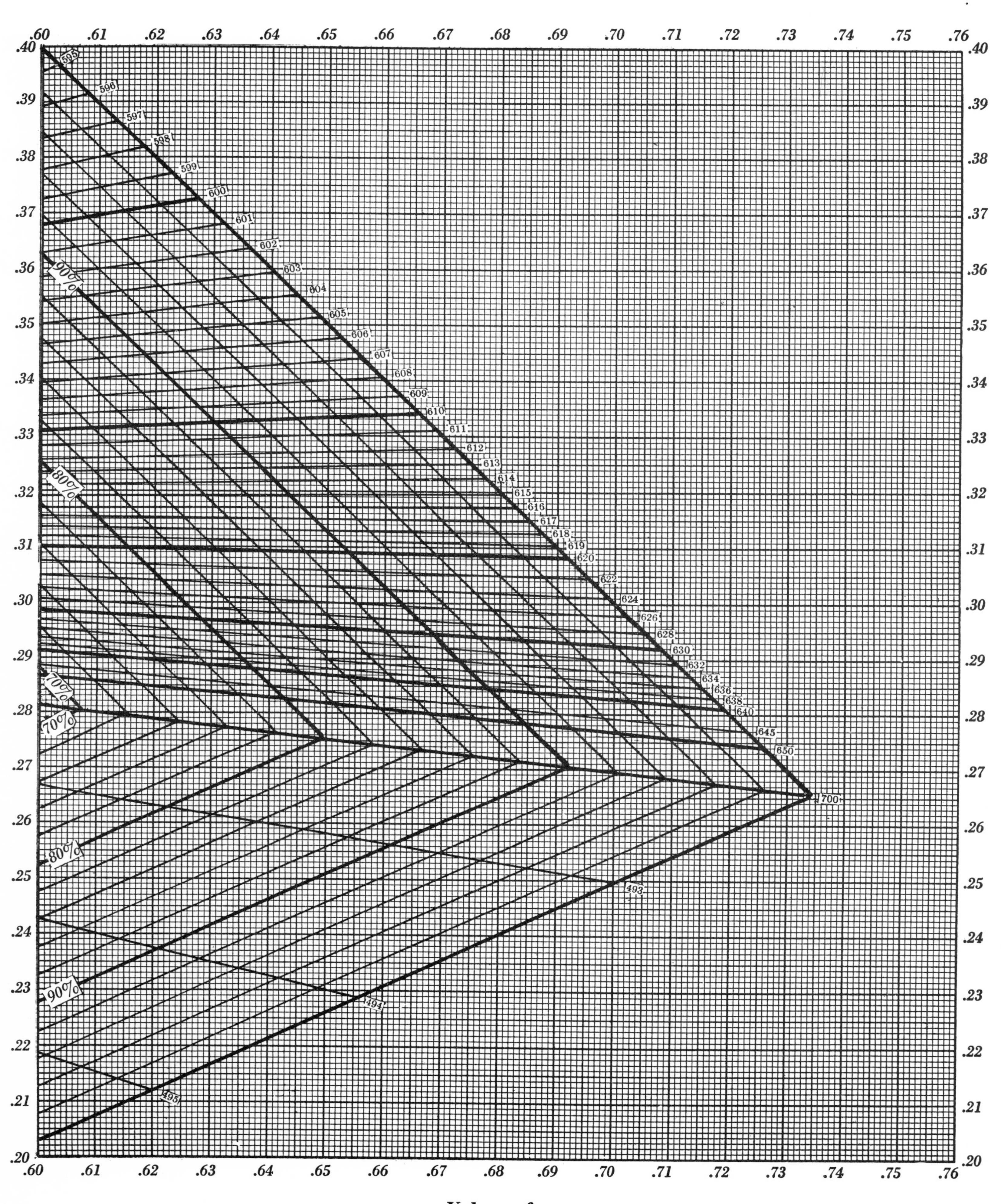

Values of y
Values of x
.60
.61
.62
.63
.64
.65
.66
.67
.68
.69
.70
.71
.72
.73
.74
.75
.76
.20
.21
.22
.23
.24
.25
.26
.27
.28
.29
.30
.31
.32
.33
.34
.35
.36
.37
.38
.39
.40
90%
80%
70%
595
596
597
598
599
600
601
602
603
604
605
606
607
608
609
610
611
612
613
614
615
616
617
618
619
620
622
624
626
628
630
632
634
636
638
640
645
650
700
493
494
495

Chart No. 23

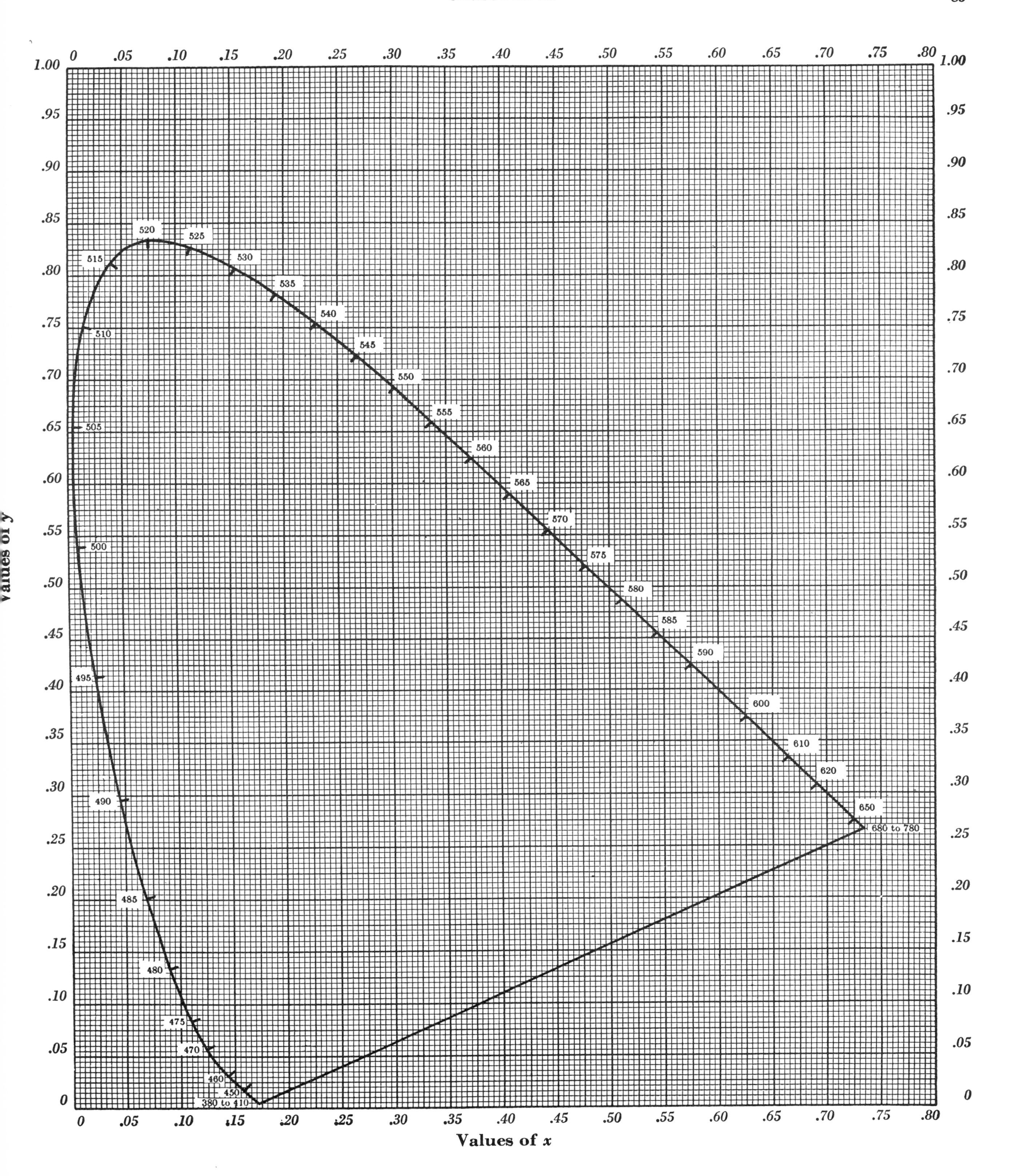

Values of x
Values of y
0
.05
.10
.15
.20
.25
.30
.35
.40
.45
.50
.55
.60
.65
.70
.75
.80
.85
.90
.95
1.00
380 to 410
450
460
470
475
480
485
490
495
500
505
510
515
520
525
530
535
540
545
550
555
560
565
570
575
580
585
590
600
610
620
650
680 to 780

INDEX